Reliability Analysis for En

D0304044

e   9035

Reliability Analysis for Engineers

An Introduction

ROGER D. LEITCH

OXFORD   NEW YORK   MELBOURNE

OXFORD UNIVERSITY PRESS

1995

Oxford University Press, Walton Street, Oxford OX2 6DP
Oxford New York
Athens Auckland Bangkok Bombay
Calcutta Cape Town Dar es Salaam Delhi
Florence Hong Kong Istanbul Karachi
Kuala Lumpur Madras Madrid Melbourne
Mexico City Nairobi Paris Singapore
Taipei Tokyo Toronto
and associated companies in
Berlin Ibadan

Oxford is a trade mark of Oxford University Press

Published in the United States
by Oxford University Press Inc., New York

A catalogue record for this book is available from the British Library

Library of Congress Cataloging in Publication Data
Leitch, Roger D.
Reliability analysis for engineers / Roger D. Leitch.
Includes index.
1. Reliability (Engineering)—Statistical methods. I. Title.
TA169.L453 1995   620'.00452'072–dc20   94–43731
ISBN 0 19 856372 8 (Hbk)
ISBN 0 19 856371 X (Pbk)

Typeset by Colset Pte Ltd, Singapore.
Printed in Great Britain by
Bookcraft (Bath) Ltd
Midsomer Norton, Avon.

I dedicate this book to
Kathy

Foreword

Professor Peter Hill
*Director of the Centre for Integrated Design, Coventry University
and The Design Consultant for the Engineering and Physical
Sciences Research Council, UK*

Gone are the days when any industry can expect to sell its products on the basis of clever functions or new technology alone. Customers now expect those clever functions to be repeatable to order over what they perceive as a reasonable length of life for the product in question. Additionally, the word 'customer' is now expanded to encompass domestic consumers, industrial customers, and those who procure sophisticated defence systems. The wise designer now takes into account all who will have to install, commission, use, repair, service, and dispose of artefacts; and for all who may come into contact with them incidentally. Thus products are designed for, in effect, a wide range of users, and those users may be virtually anybody.

If the problems of repeatability of performance by man-made devices are coupled with the problematical human and physical environments in which they must exist and operate on demand, we find new concepts emerging and new demands being made on managers and designers. Many of those concepts and demands are embodied in the single word *reliability*.

Roger Leitch has always considered reliability to be essentially an engineering problem, but one which benefits by the probing and manipulation afforded to it by mathematical modelling. His approach is realistic but rigorous in this book, as it is in his research, his consultancy, and his teaching. In his willingness to analyse and model real situations he is not one to seek spurious accuracy or believe his results without validation. I learned this when working with him on early technical risk assessment projects to help HM Government to make decisions on purchase of multi-million-pound electromechanical systems. He is the engineer's mathematician and he is the mathematician's mathematician. He is also a visionary with common sense.

I believe his approach in this book will support that view and that he will have a ready appeal to all who are involved in product introduction and reliability achievement.

Preface

After several years as a reliability practitioner, I realized that most books on the subject are either written by engineers for engineers or by statisticians for statisticians, and as a statistician who has presented many successful lecture courses to engineers, I decided to try writing a book on the statistical techniques for engineers. It is unfortunate that statisticians in academia are generally not good at communicating their subject to engineers and other users, with the result that in a good many instances statistics is either used badly or not used at all. I have attempted to correct this situation by presenting a minimum of theory, with the applications presented first, and the theory that backs it up in later chapters. For this reason I believe that practising engineers with an interest in reliability may find this book useful, as well as undergraduates and even postgraduates who are studying the subject.

The analytic techniques presented are those that are commonly used during development in the order that the project manager would normally come across them during a project. The standards in common use, BSI 5760 and the defence standard 00-41 were the source of guidance when deciding which techniques should be dealt with, but it must be emphasized that this book does not deal with all techniques, and some hard decisions had to be made when the space constraints imposed by the publisher were realized. I believed then and still believe now that it is better to treat fewer topics thoroughly than to discuss more methods at little or no depth. However, some more advanced techniques are discussed in the final chapter, and two aspects, Markov modelling and Bayesian statistics are presented in more detail. The interested reader will have to do some more reading if he or she is to really get to grips with these topics.

Chapter 1 presents the basic definitions and ideas, including the exponential distribution and the idea of a constant failure rate, Reliability Block Diagrams, and parts count and parts stress. The latter is not dealt with in any detail, as an extensive engineering input is required, and a separate book the size of this one could be written about it. Chapter 2 is on statistics and probability, and presents the basics needed to deal with the topics presented in Chapter 1, and could be omitted, at least at a first reading. Chapter 3 describes FMECA and FTA, as analyses to be done during design, or very early in the project, possibly before any prototypes are built. Chapter 4 deals with reliability data, generated, or possibly generated, during development and acceptance testing. Chapter 5 is a second chapter on statistics, and extends Chapter 2 into significance, confidence, goodness of fit tests, and linear regression, and like Chapter 2 could be omitted at a first reading, or even omitted altogether. Chapter 6 is on advanced techniques.

During the writing of this book, I have had help from a number of people, and this is the place to thank them for their support and patience. I typed the script myself, straight into a word processor (a practice I can recommend), and so there is no poor secretary who had to interpret my scrawl. However, I did impose on a number of colleagues and friends, especially Stewart Uttley and Peter Hills, who read and commented on what I wrote, and to them I am very grateful. Peter in particular deserves a further mention, as having asked him to read the script, I went further and asked him to write the foreword, for which he deserves particular thanks. Lastly I want to thank my wife, Kathy, who shared my sense of achievement when I signed the contract with OUP, and supported me at the start of the project. However, even her determination and courage could not prevent the inevitable, and she cannot be with me to share my satisfaction on completing it. It is to her that I dedicate the book.

Faringdon, Oxon   R. D. L.
September 1994

Contents

Initial definitions

1.1 Introduction

Reliability is that ephemeral property of an item or service that we all
desire, but all too often find is missing. Ask the man in the street what
he understands by reliability and the answer will be that it is the ability
to do a job, and do it well, and to keep on doing it, or some similar
qualitative statement. There will often be confusion between reliability
and quality or availability. Rarely will the response resemble the quantitative definitions that appear in so many texts and standards, a version
of which is given below. There are many people, engineers and other
numerate individuals among them, who know how fast their car can go,
and who know its petrol consumption, and have an opinion about the
reliability or otherwise of the vehicle, but have no idea of its mean
distance between failure. Our hypothetical subject will almost certainly
find it easier to give examples of reliability, or lack of it, although the
reputation of the items quoted will frequently be the result of a single
experience, particularly if it is a bad one. Some systems are very commonly held up as prime examples of reliability. Nineteenth-century
engineering, for example, has a reputation for reliability that until
recently was hard to beat (they knew how to do it in those days!), but
which is due more to the durability of some of the engineering
achievements. (These are the survivors. According to Petroski (1985) a
large proportion of bridges that were built in the last century fell down.
See also the reference to Francis Webb, a nineteenth-century locomotive
engineer, by Pile (1979).

For a number of reasons, it is necessary to have a scientific, quantitative, definition of reliability. There is some debate in the reliability
community concerning the value of measuring, and predicting reliability,
particularly as it is a perceived property. The opposition to the practice
of quantifying reliability is small, however, if somewhat vocal, and
pleads for a return to good engineering practices. It must never be
forgotten that high reliability depends on good engineering, and that no
amount of data collection and analysis will improve reliability by itself.
A parallel lies in a simple example that is not to do with reliability but
is close to the hearts of most of us: knowing how much you spend and
what you spend it on will not improve your bank balance or decrease

the size of your overdraft. On the other hand, it is a frequently quoted maxim that you cannot manage what you cannot measure, and when producing a complex system, in a project that may take several years, high reliability will only be built into the product if the project is well managed. To return to the financial example above, if you do not know how much you spend, you will have no chance of controlling your budget and very little chance of saving any money. Indeed, in the business world, no project would get approval to proceed without a cost analysis prediction showing that it would eventually make a profit; this is despite the fact that most of the models used are poor, and that experience tells us that the predictions are frequently wrong, usually optimistic.

There are other reasons for measuring reliability: for example, to gain assurance when purchasing goods (this is particularly true when a large, complex organization such as a railway company or the armed forces is procuring equipment), for contractual purposes, to optimize maintenance policies, or to predict running costs and spares holdings of complex equipment.

The purpose of this chapter is to introduce some reliability terms, and some of the implications of the basic formulae connecting them. The mathematics in this chapter is kept to a minimum, and is expanded further in Chapter 2.

1.2 Definitions

The reliability of a product is the measure of its ability to perform its function, when required, for a specified time, in a particular environment. It is measured as a probability.

The definition contains four important terms:

> function
> environment
> time
> probability

These terms will be discussed in more detail.

1.2.1 Function

It may seem an easy matter to decide on whether or not an item is functioning correctly, but this simple aspect of reliability has caused more distress in the relationship between customer and supplier than any other. It is essential that the ground rules for sentencing anomalies or defects are made clear at the start of any project or negotiations. A failure may be defined as any defect that causes an unplanned maintenance action, or one that impairs operation. It may be that there is more than one level of performance (war or peace, for example, for a military system), and the reliability at each level has to be defined and monitored.

For example, the function of a domestic fridge is to keep food and drink cold. What if the light fails? It still performs its primary function, and few of us would discard a fridge because of such a failure, but it would cause dissatisfaction to the owner, particularly if it was new. This demonstrates two levels of failure in a simple system. How much more difficult is the problem of defining failure when a more complex system, such as a central heating system, or a vehicle, is being considered.

When the definitions of function and failure are being formulated at the start of a development project for a complex piece of equipment, the project manager and his team must foresee the effects of these definitions on possible failure scenarios that may occur later in the project.

Generally, failure is caused when a component fails either catastrophically, or by drift. The former is when something breaks suddenly, as when a light bulb goes, or a tyre punctures. The latter is due to some ageing process, such as wear or fatigue, and causes the parameters of interest of the components concerned to change, and go out of tolerance. These two causes are related, as fatigue, for example, may cause a catastrophic failure if not detected and dealt with by appropriate preventative maintenance. Ageing may be considered a durability rather than a reliability problem, but from the user's viewpoint the perceived reliability is important rather than what to him may be a nit-picking discussion of definitions.

1.2.2 Environment

The reliability of any equipment is very dependent on its surroundings, or environment. This not only includes the climate, which is generally the first thing that springs to mind when environment is mentioned, but for example,

packaging, transportation, and storage

installation

the user

maintenance resources available

dust, chemical, and other pollutants

and others. Examining the environment is an essential part of specifying and estimating reliability. Not doing so makes the consideration of reliability meaningless.

It is worth saying a few words about the above aspects of the environment. Packaging, transportation, and storage are extremely important if the equipment is to reach the user in prime condition. Indeed, for many systems the worst stresses are experienced between factory and installation. Even equipment that is regularly used may not be stored in optimum conditions, and this can include domestic equipment. How many domestic appliances, which could be thought of as delicate, are stored in a garage or garden shed, where the temperature may vary enormously, and the humidity may be very high?

Installation can also cause extremes of stress. During installation, equipment is probably being moved, may not be the right way up, could be opened allowing the ingress of dirt and other pollutants, and is subject to many other stresses.

The user is a prime example of sources of failure in a system, and it is essential that the equipment designers take his or her fallibility into account. They must consider the user's training, intelligence, morale, and interest in the task in hand, as well as ensuring that the potential for mistakes during operation is minimized, by appropriate consideration of the design of the controls and instrument panels. Good user testing is essential to ensure that this aspect is appropriately addressed. A good example of this has been the design of video cassette recorders and their programming in particular. A few years ago they were very user unfriendly, the programming was awkward, and the manuals poorly written. It was evident that any testing done was using subjects that were fairly familiar with programming and electronic equipment, rather than using subjects who were more representative of the population in general. The office cleaner, the managing director's wife, and a security guard would be ideal members of such a test team. Things have been getting better in the last few years, and manuals and programming techniques have improved, but it is still the case that the user is too frequently not properly considered in the design of complex equipment.

Concerning maintenance and maintenance resources, the author has been frequently asked in the past if it is right that the reliability of equipment is stated assuming that a certain amount of preventative maintenance is done. The answer is yes, as the preventative maintenance is part of the design. From asking his students over many years, he has come to the conclusion that the reliability in one year of the modern car is between 90 per cent and 95 per cent, when failure is defined as not being able to drive the car any further and having to be towed away for repair. This is a very broad brush approach, and begs any questions concerning age, make of car, etc. The vast majority of us do, or get done, the preventative maintenance necessary (change the oil, clean the plugs and points, etc.). So this figure, of 90 per cent to 95 per cent, assumes the preventative maintenance has been done. In any case, the maintenance resources available will influence the desired reliability. You may be prepared to accept one level of reliability when driving to and from work, but if you were driving across the Sahara Desert you would probably insist on a much higher value.

The presence of pollutants such as dust, salinity if near or on the sea, etc., clearly will affect the ability of the equipment to perform its function and continue to do so, and no more need be said about this aspect.

1.2.3 Time

Basic {

Reliability decreases with time, in the sense that as the task duration increases, there is a greater probability of failure. However, the duration of the task, often called the mission, may not be measured only in units of time, but could be in terms of distance travelled, as for a vehicle, or duty cycles, as for a pile driver, for example, or some combination of these or other parameters. The remainder of the book will generally use time, but only in a generic sense, and the reader must remember that other units may have to be substituted.

An item will also suffer from ageing, as mentioned earlier, which will alter its ability to perform its function, and the ageing processes will also be a function of time and/or distance, and/or cycles, etc.

For example, a car on a long journey is more likely to fail than one on a short journey. There is simply more opportunity for it to do so. Also, a very old car, which has completed a high mileage, will be less capable of completing a standard journey than a relatively new one. The reader may have experienced a degree of foreboding when faced with a long journey in an 'old banger'.

Exercises 1.1

1.1.1 For each of the following systems, describe briefly (i) the environment, and (ii) the mission you would expect them to experience.

(a) A domestic freezer

(b) A fire extinguisher (for a warehouse)

(c) A floodlight for a sports ground

1.1.2 It is often necessary to categorize failures according to how serious they are for the user. For example, consider the following failures of a domestic car:

brake failure

bulb in courtesy light fails

silencer falls off

fuel pump fails

puncture

puncture, but cannot change the wheel

Clearly some of these failures are of a more trivial nature than others. Devise a categorization of domestic car failures, and put each of the above failures in the appropriate category.

1.1.3   How would you measure time in each of the examples in Question 1.1? How would you define a mission, and how long would you expect the equipment to last?

1.1.4   What are the functions of a fire alarm system? Consider the cases when there is a fire, and also when there is no fire. What values would you consider reasonable for the reliability? What considerations led to your conclusions?

1.3 Probability

Reliability is measured as a probability, and what is more, a probability that changes with time, and so we are in to the realms of statistics and statistical analysis. This section is devoted to presenting the definitions of the statistical terms needed to appreciate and discuss reliability, and the statistical analysis is kept to a minimum.

1.3.1 Reliability with time

Suppose a large number of simple items are put on test together. Even though they are nominally identical, same manufacturer, same design,

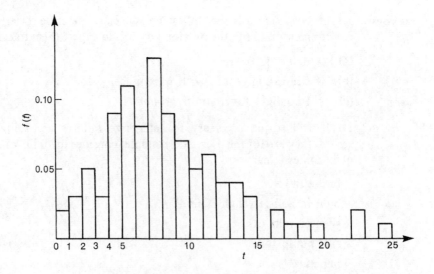

Fig. 1.1 Histogram of lifetimes.

same batch number, etc., and the environment is the same for all of them, they will not all fail at the same time. It is said that the lifetimes are distributed. If these data, the times to failure of all the units, were given to a statistician, the first thing that might be done with it would be to construct the histogram shown in Fig. 1.1. The horizontal axis is time, the ageing parameter, or mission duration, and the vertical axis can be a proportion, as it is here, or a count. Time is divided up into suitable intervals by the values $t_0 = 0, t_1, t_2, \ldots$, and f_i is the proportion of items that failed in the ith interval, that is the one between t_{i-1} and t_i. If n_i is the number of items that failed in the ith interval, then

$$f_i = \frac{n_i}{N}$$

where N is the initial total number of items on test. So from the histogram it can be seen that approximately 0.02, or 2 per cent of these items failed between times zero and one, or in the first hour, if we assume that time is measured in hours, about 3 per cent failed in the second hour, 5 per cent in the third hour, and so on.

This is not the most useful way of presenting the data from the reliability point of view, and Fig. 1.2 shows the cumulative, or ogive, curve. The height of each bar is the sum of the heights of the bars of Fig. 1.1 up to the appropriate time, i.e.

$$F_i = f_1 + f_2 + \ldots + f_i$$

$$= \sum_{j=1}^{i} f_j.$$

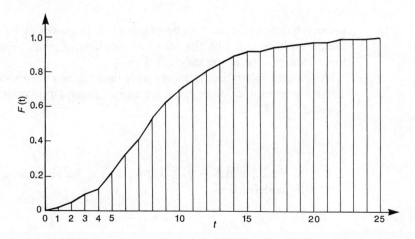

Fig. 1.2 Cumulative graph of lifetimes: the Failure Function.

So, for example,

$$F_1 = 0.02$$

$$F_2 = 0.02 + 0.03$$

$$= 0.05$$

$$= 5\%$$

$$F_3 = 0.02 + 0.03 + 0.05$$

$$= 0.10$$

$$= 10\%.$$

The interpretation of this is that 5 per cent of the items had failed by the end of the second hour, 10 per cent by the end of the third hour, etc.

This graph is of far more use than the first histogram, as it is of far greater interest to know that if, for example, the task duration is three hours, 10 per cent of the items are not going to make it, i.e. the reliability at three hours is 90 per cent. Precisely when, in that three hour mission, the 10 per cent are going to fail is not of primary interest.

To summarize,

f_i is the proportion of items that fail between times t_{i-1}

and t_i, while

F_i is the proportion that fail by time t_i.

Reliability is really what is of interest, and if F_i is the proportion that fail by t_i, then $1 - F_i$ is the proportion that still function at time t, and so

$$R_i = 1 - F_i$$

is the reliability at time t_i. The graph of R_i is shown in Fig. 1.3. (Some authors call this the survival function, S_i, the proportion of items that survive up to time t_i.)

Notice that reliability decreases with time, as was mentioned earlier. There are fewer and fewer survivors as time passes, until eventually they have all failed.

1.3.2 Failure rate

The failure rate, which is also known as the hazard rate, or force of mortality, is defined as

$$\lambda_i = \frac{f_i}{R_{i-1}}.$$

When f_i and R_i are defined from the data in the way described above

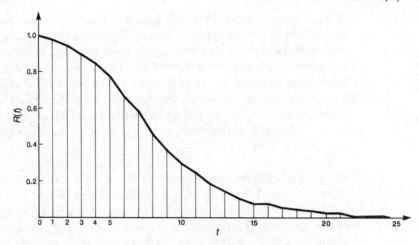

Fig. 1.3 The reliability function.

then λ_i is the probability that an item that has survived up to time t_i will fail in the next time interval. Clearly, the smaller the value of the failure rate, the more reliable the equipment for a fixed time period. The relationship between λ and R will be discussed in more detail in the next chapter.

1.3.3 The bathtub curve

For many items, a graph of failure rate with time often looks qualitatively like that shown in Fig. 1.4. Because of the shape of the graph, it is known as the bathtub curve.

Consider first a simple, non-repairable component. In a large batch

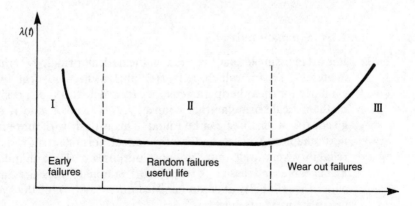

Fig. 1.4 Qualitative behaviour of the failure function: the bathtub curve.

of such components, there will be some that are faulty or weak, no matter how good the quality control, inspection procedures, etc., and these will fail early. This will manifest itself as a high initial failure rate. Once these items have been weeded out, the failure rate will stay low and approximately constant, and this period is called the useful life period. Eventually, the ageing processes that have been mentioned earlier will assert themselves, and the failure rate will increase again. This is the wear out period, and if appropriate, it may be the time to change the component before it fails.

For a more complex, repairable item the story is similar. Manufactured faults in some of the items will give an initial high failure rate. Once such faults have been dealt with the failure rate comes down and stays more or less constant. (Those readers who have been fortunate enough to own a new car will know that it has to be treated very carefully for the first few months of its life.)

It might be asked, if the item can be repaired, why the failure rate ever starts to increase? Given enough time, it will indeed settle down to a level that is approximately constant, but it goes through a number of cycles of rising and falling before that phase is reached. The first one will occur at the stage when many of the components are failing, and if just this part of the graph is considered, it looks like the bathtub curve. There is good reason for cutting the curve off at the first cycle, as when costs of purchase and repair are considered, and averaged out over the whole life, this average reaches a minimum value at the start of the wear out period. This is because the cost of a spare part, and of fitting it, is higher than the initial cost of the same item to the producer. (For example, building a car out of spare parts would cost several times more than buying it new.) Carter (1986), does a detailed analysis of this phenomenon, including a simple numerical example.

1.3.4 Summary measures

It is often assumed that there is a mathematical formula $f(t)$ that approximates f_i, or for which f_i is $f(t)$ plus some statistical 'noise', and similarly for the other parameters, F, R, and λ. It is then possible to find mathematical formulae that connect f, F, R, and λ, and if any one is given, the other three can be found. This is dealt with more fully in the next chapter. Curves and mathematical formulae may describe the reliability very well, but for most purposes a single number is more manageable and easier to deal with. Summary measures are usually either the reliability at some standard time t', $R(t')$, or the average time to, or between, failure.

$R(t')$ is the proportion of items that survive without failure to time

t', possibly allowing some maintenance. This is the most straightforward measure of reliability, and the most useful.

The mean time to failure (MTTF) for non-repairable items, or the mean time between failure (MTBF) for repairable systems is the average time the operator can use the equipment before it fails. Note that it is an average, and that some items will last longer than the MTBF, while some will not last as long. It is not a guaranteed minimum.

1.4 A model for failure rate

If a formula, $f(t)$, is assumed for the distribution of failure times, this is called a model for failure rate. Here we shall examine the consequences of assuming that the failure rate is constant in which case (as will be shown in the next chapter)

$$R(t) = e^{-\lambda t}.$$

It also turns out that in this case, *and only this case*,

$$\lambda = \frac{1}{\text{MTBF}}$$

and so

$$R(t) = \exp\left(-\frac{t}{\text{MTBF}}\right).$$

It is important to realize the consequences of this formula. Figure 1.5 shows how reliability varies with time for a military vehicle, for the two values of the MTBF of 10 battlefield days and 20 battlefield days. (A battlefield day is a twenty-four hour period during which the vehicle will have travelled a specified distance and performed some standard tasks.) Notice that the reliability drops off very quickly initially, and by the time the MTBF is reached, the reliability is about one third. This means that two vehicles out of three have needed at least one repair, and if it is the reliability of a non-repairable item that is under discussion, then by the MTTF two items out of three have failed and are lost. It is for this reason that a very long MTBF or MTTF is needed to obtain a reasonable reliability over a standard time period.

The constant failure rate model, or the exponential model as it is also called, is very popular, partly because of its simplicity and ease of use, and partly because of its applicability to complex systems, particularly in their useful life period. It can be proved mathematically that given some very general conditions, for a repairable system, after an initial 'settling in period' the failure rate is approximately constant. (See for example Cox (1967).) We shall return to this point in a moment.

One property of a system with a constant failure rate is that it

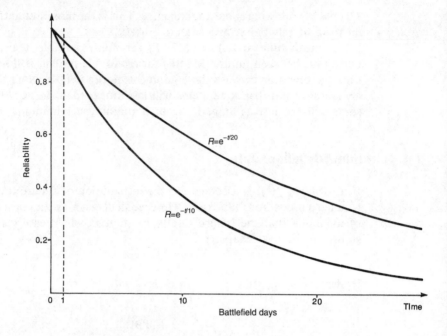

Fig. 1.5 Example of the relationship between reliability and time: a military vehicle with constant failure rate.

apparently does not age. In quantitative terms, this means that the probability of failure of such a system depends only on the period of time it has to function, and not on the age of the system. (Do you believe this about your car?) In qualitative terms, it means there appears to be no deterioration of the system until it suddenly fails. Notice that it only *apparently* does not age, and that there only *appears* to be no deterioration. It is better to say that the system has no memory.

It may seem far fetched to be asked to believe that there are such systems, but for a complex, repairable system, it will eventually settle down to a state where its components are a mix of old and new, in dynamic equilibrium. If the failure rate is averaged out over a reasonable time period and, in particular, if a fleet of systems is being considered, averaged over the fleet, the failure rate appears constant. This model is useful for predicting global parameters such as spares requirements, and system and fleet availability, but is poor at predicting local behaviour such as mission reliability for a single system.

For components, the constant failure rate model is considered acceptable for electronic components in some applications, although they are known to age. This is because in some applications, electronic components are more likely to fail through shocks from the environment, that occur randomly. The dominant failure modes for mechanical components however, are due to ageing, and an increasing failure rate model is needed to describe their behaviour. A suitable model is described later.

Exercises
1.2

1.2.1 Table 1.1 shows the relationship between the reliability, mission time (in hours), MTBF (in hours), and failure rate (failures per hour) of a number of complex systems. Assuming a constant failure rate, complete Table 1.1 by filling in the missing entries.

Table 1.1 The MTBF, failure rate, mission time, and reliability for six systems

System	1	2	3	4	5	6
MTBF		150				200
λ	0.01					
Mission	10	20	25	10	15	
R			0.95	0.99	0.8	0.9

1.2.2 Ten identical items with constant failure rate were put on test, and their failure times (in hours) were observed to be

$$10 \quad 17 \quad 25 \quad 33 \quad 34 \quad 41 \quad 48 \quad 59 \quad 72 \quad 79.$$

Calculate the MTTF, and hence the reliability at 20 hours.

1.2.3 In a similar test, 20 items with constant failure rate were tested for 50 hours, by which time 6 had failed at times

$$17 \quad 24 \quad 33 \quad 35 \quad 42 \quad 49$$

while the remaining 14 were still functioning. Calculate the MTTF and the reliability at 30 hours. (*Note*. In this case, the MTTF is the *total* time on test, divided by the *total* number of failures.)

1.2.4 In the two previous questions, give another estimate of the reliability that does not involve finding the MTTF, or the assumption of a constant failure rate.

1.5 Maintainability

Although the engineering activities involved in ensuring maintainability are very different from those needed to ensure reliability, no text on reliability is complete without at least a mention of maintainability. The graph of Fig. 1.6 shows a typical (though simplified) history of a repairable system. Notice that it spends some of its life up and running, and the remainder down being repaired. So far, we have only considered the up times. Maintainability is the study of the down times.

Observe that the down times are all different, and so are also

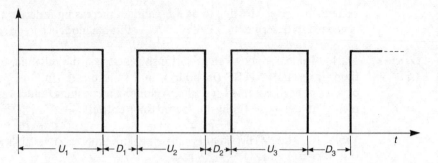

Fig. 1.6 History of a repairable system: up and down times.

distributed, and it is generally impossible to tell in advance how long a repair will take. So probability and statistics are involved in the study and definition of maintainability. The definition is

Maintainability is the probability that a failed system is restored to the functioning state, in a given time and in a stated environment, which will include the maintenance resources available.

It is denoted $M(t)$. It is possible to analyse maintenance times in a similar way to that used for failure times. In this case m is used for the distribution of maintenance times, corresponding to f, and μ is used for maintenance rate, corresponding to λ for failure rate. There will not be a detailed discussion of the statistics of maintenance times as there was for failure times. The concept of maintenance rate is rarely used except in Markov modelling (see Chapter 6).

Figure 1.7 shows a hypothetical distribution of maintenance times. The mean time to repair (MTTR) is shown, but just stating a mean is not always useful, as was shown earlier when considering the MTTF and MTBF. When considering maintainability, it is advisable that one also examines a 'worst case', in this case the 95th percentile. So, for example, a specification may state the MTTR, but should also state that 95 per cent of all repairs will be completed by a specified time, as shown in the Fig. 1.7.

Exercises 1.3

1.3.1 State the maintainability requirements for the following repair actions:

(a) changing the wheel on a car

(b) changing the fuse in an electric plug

(c) renewing a tap washer

In each case analyse the task (i.e. decide what has to be done) in order to obtain your requirement. Write down some of the difficulties that may be met when trying to carry out these activities.

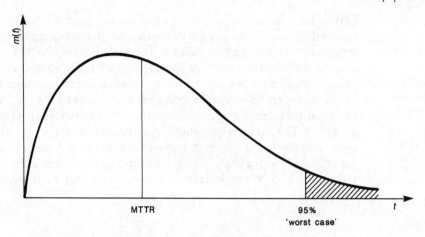

Fig. 1.7 An example of the distribution of maintenance times.

1.3.2 In the above question, state the environment that you have assumed, and the resources available. What assumptions have you made concerning the individuals involved (training etc.). How much could all your assumptions affect your requirement?

1.6 Availability

In many situations, the availability of a repairable system is of primary importance to the user, rather than the reliability itself. The definition is:

The availability of a system is the probability that the system is functioning at time t.

It is denoted $A(t)$. Initially, it may be reasonable to assume that $A(0)$ is 100 per cent, although with equipment that has just come out of storage, $A(t)$ may well be below this. However, given enough time, a steady state value will be reached, which is the ratio of up time to total time, or if a fleet is involved, the proportion of the fleet that is functioning at any time. The steady state value, A, is given by

$$A = \left(\frac{\text{MTBF}}{\text{MTBF} + \text{MTTR}} \right) \qquad (1.1)$$

The implication of this formula is that a high availability can be obtained either by increasing the MTBF, and hence the reliability, or improving the maintainability by decreasing the MTTR.

It must be observed that from the user's point of view, the MTTR must include the time it takes the repair team to arrive, and possibly send for spares (the logistic delay). The designer can affect the reliability, and that part of the maintainability which includes the diagnosis of the fault, the removal of the faulty part, its adjustment or repair, replacing it, and testing it, but the logistic delay is the responsibility of the user. The average time taken to perform the engineering tasks without any logistic delay is sometimes referred to as the mean active repair time (MART), or the mean active corrective maintenance time (MACMT). The use of this figure in eqn 1.1 gives a value of the availability known as the intrinsic availability. It is often quoted by manufacturers when advertising their products (as they cannot know the value of the logistic delay). This value is the best that can be achieved, and in practice is not achievable, unless there is a repair team accompanying and dedicated to each equipment.

1.7 Reliability systems modelling

As was said earlier, there is some debate in the reliability community about the value of modelling reliability, and the need for predictions. Many predictive techniques are not accurate, and often demand a degree of effort and skill that does not always appear to be cost effective. Often this is due to a misunderstanding of the assumptions made, and their effects on the model. These criticisms come from those engineers who are not interested in theoretical techniques. However, being able to model a system, and perform some analysis in itself gives some assurance that the designers are applying themselves to ensuring a reliable product, and of course the numerical prediction gives further assurance. This is in addition to being able to play 'what if' games when deciding on types and degrees of redundancy, spares requirements, maintenance policies, etc.

There are many definitions of model, one of which is:

A model is a set of rules or equations that describe the behaviour of a system.

Models are used as an aid to understanding the structure of a complex system, for prediction, and as an aid to decision making. The reliability structure of a system can be modelled in principle by the use of structure functions and Boolean algebra, although this approach is rarely used directly in practice, but all models must reflect the logical structure of the system. The analyst need not necessarily be aware of this background to the model, however, particularly if he has the appropriate computer support.

As the project proceeds through the development cycle, different modelling activities are required. It must also be remembered why the model is needed (i.e. what questions are being asked), what the model boundaries are to be, and what assumptions are being made. Any model must compromise between the detailed aspects of the system to be modelled and the need to produce an answer in a reasonable time and to reasonable cost, but the effects of the assumptions on the conclusions must be considered. As the project proceeds into the later stages of development, the results of the model can be compared with real data (validation), and the experience gained used in subsequent projects. A weakness of any model is the need for good data, and any decisions made as a result of modelling, particularly in the early part of development, must take account of the sensitivity of the results to possible variations in the data. It must also be remembered that reliability is an engineering activity, and that data in the form of engineering reports and expertise must also be an input into the decision process.

1.8 Reliability block diagrams

1.8.1 Introduction

Reliability Block Diagrams (RBDs) are a simple model that can be used at all stages of the development of a product, and give a pictorial representation of the reliability structure. For example, an ignition system of a car can be thought to consist of the points, distributor, cables, and battery. When starting the car, all these components are needed, and this is represented by the RBD in Fig. 1.8.

Once the car is running the electrical power is normally taken from the generator, although in an emergency the car will function by taking power from the battery (for sufficient time to get to help if the drive belt breaks, for example). There is a degree of redundancy in the system, and this is represented by the RBD in Fig. 1.9.

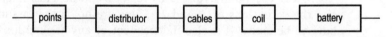

Fig. 1.8 The RBD of the ignition system of a car when starting.

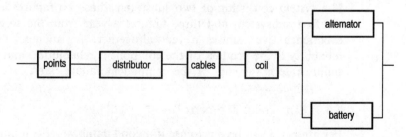

Fig. 1.9 The RBD of the ignition system of a car when running.

When modelling using RBDs, there are essentially two situations, *series*, where all the components are needed, and *redundancy*, when there is some standby facility, or alternative means of keeping the system running.

1.8.2 Series systems

Series systems are those in which all the components must function for the system to function, and are represented by the diagram in Fig. 1.10.

Fig. 1.10 Series system.

In this case, if all the failures are independent, then

$$R_s = (R_1 \times R_2 \times R_3 \times \ldots \times R_n) \tag{1.2}$$

where R_s is the system reliability and R_i is the reliability of the ith component, for $i = 1, 2, 3, \ldots, n$. By *independent* is meant that knowing whether or not some of the components are functioning or failed does not make it any more or any less likely that the remainder are in a similar state. A common reason for lack of independence are common cause or common mode failures. The above equation comes from a basic statistical law which is discussed in Chapter 2, and is more likely to hold for electronic systems than for mechanical ones.

Given eqn 1.2, it can be shown that

$$\lambda_s = \lambda_1 + \lambda_2 + \lambda_3 + \ldots + \lambda_n \tag{1.3}$$

where λ is the failure rate. This relationship depends only on eqn 1.2, and the definitions of R and λ. There are no further assumptions that have to be made. It is also independent of any assumptions concerning constant failure rate, although if the failure rate is not constant, the mathematics soon becomes untenable except in relatively simple cases. Equation 1.3 must be true, if one thinks about it for a few moments. If a system consisting of two subsystems has two failures a year from the first subsystem and three failures a year from the second, it will experience five failures a year altogether. In summary, the system reliability is the product of the component reliabilities, and the system failure rate is the sum of the component failure rates.

1.8.3 Parts count and parts stress

Equation 1.3 can be put to use if a good database of component failure rates is available, and this technique (of summing component failure rates to predict the system failure rate) is called *parts count*. This is a technique that has been in use for a number of years, particularly for electronic systems, and can be applied very early in the design phase of a project, before any hardware is built or data has been generated. There are a number of databases available, one of the better known being MIL-HDBK 217. It assumes that the failure rates are constant.

The basic equation for parts count is

$$\lambda_{\text{equipment}} = \sum_{i=1}^{n} N_i (\lambda_G \pi_Q)_i$$

where:

$\lambda_{\text{equipment}}$ = required equipment failure rate in failures per million hours

λ_G = generic failure rate for the ith part in failures per million hours

π_Q = quality factor for the ith part

N_i = quantity of the ith part

n = number of parts

The values of the failure rates and quality factors are found in tables in MIL-HDBK 217. The failure rate also takes into account the environment in which the equipment is to operate.

In practice, this technique has not been found to give accurate results. In order to improve the accuracy, the parts stress technique was developed. The failure rates, assumed constant, are still summed, as with parts count, but the derivation of the component failure rates uses a model that takes into account far more factors than just the quality and the equipment environment. In general, this is much more complicated than parts count, and only two examples are given here for illustrative purpose only. For more information on the use of this technique, the reader must consult the handbook.

For ROMs and PROMs, for example, the component failure rate is given by:

$$\lambda_D = \pi_Q (C_1 \pi_T \pi_V + C_2 \pi_E) \pi_L$$

where:

λ_D is the required failure rate;

π_Q is the quality factor;

π_V is the voltage stress derating factor;

π_T is the temperature factor;

π_E is the application environmental factor;

π_L is the device learning factor;

C_1 is the device complexity failure rate;

C_2 is the package complexity failure rate.

The values for these parameters are tabulated in the handbook, and can be found once sufficient information is known about the design. Notice that this equation assumes that a failure either comes from the chip itself, or from the packaging. The equation is derived partly from known physical principles, and partly from curve fitting to data.

For resistors, the equation is:

$$\lambda_R = \lambda_B \pi_Q \pi_R \pi_E$$

where:

λ_R is the required failure rate;

π_Q is the quality factor;

π_R is the resistance factor;

π_E is an application factor;

λ_B is given by:

$$\lambda_B = A \exp B \left(\left(\frac{T+273}{N_T} \right)^G + \left(\frac{S}{N_S} \frac{T+273}{273} \right) J^H \right)$$

where:

A, B, G, H, and J are constants;

T is the temperature in degrees Celsius;

N_T is a temperature constant;

S is the electrical stress;

N_S is a stress constant.

Again, this equation is a mix of physics and curve fitting.

Despite the effort involved in producing the data and the techniques, the results are still not very accurate, and must be used with caution, particularly when large complex systems are involved. There are a number of reasons for this lack of accuracy, among which are:

• transferring data from one application to another;

• statistical variability in the data;

• inappropriate physical models (for example, Arrhenius' equation relating failure rate to temperature).

However, it must be remembered that the predicted figure is not a promise or guarantee that the final system will have that reliability when it is finally built. It can only be used as a guide, as an aid to decision making and to gain some assurance, and once the decision is made to go ahead with developing a particular design, the parts count/parts stress prediction can be put aside as data from trials and tests are generated. On the other hand, it is easy to do, and with computer support need cost very little. (Predicting anything has never been an easy task throughout the history of the human race, and those who were thought to be able to prophecy the future have been sought after and revered. Most of us cannot predict how much money we should take away with us on a two week holiday, much less predict the behaviour of a complex system, possibly several years before it is built and brought into service.)

Finally, and this applies to any predictive technique, actually doing the analysis necessary to produce the prediction gives the designer and others insight into the design, forces them to think about reliability at an early stage of the project, and imposes a discipline which otherwise may not be there, and if those responsible for the design, development and production cannot give a quantitative assurance of the final reliability, then there is no assurance, and one could be forgiven for believing little chance, that a high reliability will be achieved.

1.8.4 Redundant systems

Systems with redundancy are often called *standby* or *parallel* systems, and are represented by a diagram like that in Fig. 1.11.

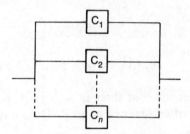

Fig. 1.11 Redundant system.

The situation is very much more complex than that of series systems, and there are no simple equations like eqn 1.2 or eqn 1.3. For example, Fig. 1.11 could represent a two engined aircraft able to fly on one engine. In this case it can be shown that

$$R_s = 1 - (1 - R_1)(1 - R_2)$$

and the equation for the failure rate is even more fearful. Note that this equation does not take into account the added stress on the surviving engine should one of them fail.

Redundancy may be *active* or *passive*, i.e. the redundant items may be switched on when the system is being used, as in the case of a multi-engined aircraft that can fly if some of its engines fail, or they may be left until needed, as with a spare tyre on a car. The latter case is often

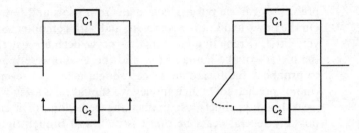

Fig. 1.12 Conventions for showing passive redundancy.

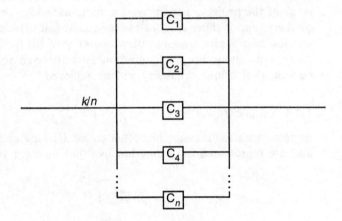

Fig. 1.13 Diagram showing k-out-of-n redundancy.

called standby, but there is no agreed use of this term. Passive redundancy is often denoted in an RBD by one of the conventions shown in Fig. 1.12.

Redundancy is also classified on a k-out-of-n basis, i.e. the system consists of n components, of which any k functioning will ensure the system functioning. The system reliability in this case is calculated by use of the binomial distribution, (see Chapter 2), although when the component reliabilities are unequal the resulting expression becomes very unwieldy. A k-out-of-n system is indicated as shown in Fig. 1.13.

For example, Fig. 1.14 shows the engines of a four engined aircraft that can fly on one engine. In different circumstances, (for a different load or journey, for example), it may need at least two engines to carry out its task, and this is shown in Fig. 1.15. In different circumstances again, it may need at least one engine on each side, and this is shown in Fig. 1.16. A totally different example, which has been mentioned in the text, is that of the tyres of a car, which if the spare is included, are a 4-out-of-5 cold redundant system, and this is shown in Fig. 1.17.

Reliability block diagrams can also be used to predict availability, as

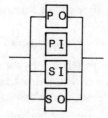

Fig. 1.14 An example of redundancy: a four-engined aircraft that can function on one engine.

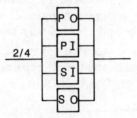

Fig. 1.15 An example of partial redundancy: A four-engined aircraft needing at least two engines to function.

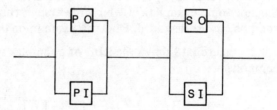

Fig. 1.16 Example of mixed series and redundant system: a four-engined aircraft needing at least one engine on each side.

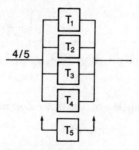

Fig. 1.17 Example of passive redundancy: a car carrying a spare tyre.

the probabilities used in the equations could equally well be availabilities as reliabilities.

Points to watch when considering an RBD include:

- More than one RBD may be needed to show different phases of system use (as with starting the car and running it once started), and to show different functions and failure modes.

- There is no account taken of the added stress survivors may experience when some components in a redundant system fail.

- In the case of passive redundancy it is difficult, if not impossible, to take account of the possibility of a switching failure (a car may carry a spare tyre, but there is a chance that the wheel cannot be changed when the puncture happens), and of the possibility of a passive failure (the spare wheel may be flat).

- There is no account taken of different maintenance and repair policies, availabilities of repair teams and spares etc., when using RBDs to predict availability.

Many of these difficulties can be overcome by the use of other techniques such as simulation, fault tree analysis and Markov analysis.

Exercises 1.4

1.4.1 Reliability block diagram modelling assumes that the components are either functioning or failed, i.e. there is no degraded state for them. Suggest some examples in which the system can function, even if the components are degraded. Does this assumption matter in your example?

1.4.2 Figure 1.18 shows the RBD of a complex system made up of four components.

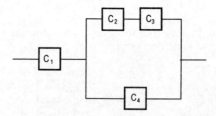

Fig. 1.18 A complex system of four components.

The reliabilities of the four components are:

component	1	2	3	4
reliability (%)	95	85	90	98

Calculate the reliability of the system.

1.4.3 It is decided to improve the reliability of the system by putting in more redundancy. In this case it can either be at system level, when a second system is kept in standby, or at component level, when each component is replaced by two identical components in standby. Calculate the reliabilities in each of these cases using the data of the previous question.

1.4.4   In Question 1.4.2, if the reliabilities are measured over a ten hour mission, calculate the MTTF of each component assuming a constant failure rate for each. Now calculate the reliability of the system as a function of time.

1.4.5   A special heavy vehicle has four identical axles each of which has ten identical wheels, five on each side. Each axle can function if it has at least four unpunctured tyres on each side. In addition there are two spare wheels carried on each vehicle.

Draw the RBD of this system, indicating the types of redundancies present (active or passive, k-out-of-n).

1.4.6   A fire alarm consists of a detector that is connected to an alarm bell. The functions of the detector are to emit a sign if a fire is present, and not to emit a signal if no fire is present. The reliabilities of the detector are believed to be as follows:

The probability of emitting a signal when a fire is present is 99.5 per cent.

The probability of emitting a signal when a fire is not present, during one year, is 2 per cent.

There is concern that the detector is not sufficiently reliable at detecting fires, and an attempt to improve the reliability is made by putting in two detectors, connected through a logical OR gate, so that a signal will be transmitted to the alarm bell if either of the detectors sends out a signal. This would clearly degrade the false alarm reliability, as there is now a greater probability of a false alarm. (This could have economic implications with loss of production if the alarm system was in a factory, for example. In any case it could affect morale, and if there were too many false alarms, when there was a real fire the alarm would probably be ignored.) Draw the RBDs of the system for the two functions, detect a fire and no false alarms.

The specification for the detection part of the system is that it should be 99.9 per cent reliable for both its functions. By using as many detectors as you wish, with logical AND as well as logical OR gates, design a system that meets this specification.

1.9 An introduction to Monte-Carlo simulation

Some systems are too complex to analyse easily by analytic methods. In cases like these, Monte-Carlo simulation is frequently used, as it is very versatile, and can be used to examine different maintenance policies, design solutions, spares holdings, and many other parameters that affect reliability, availability, or other measures of effectiveness.

Consider a component that is said to be 95 per cent reliable in a given situation. In principle, this statement comes from a trial that has been carried out, in which a number of the components have been tested, and 5 per cent of them failed while 95 per cent of them functioned successfully. A report of the trial would be full of information that is not directly relevant to the issue at hand, such as the weather during the trial, the manufacturer or manufacturers of the components, etc., while the information of interest, namely the data, would probably be confined to an annex. This would simply be a record of the success or failure of each component, and in essence would consist of a list of Ss or Fs for each component, S for success and F for failure. There may have been 20 trials with one failure, 100 trials with 5 failures or 1000 trials with 50 failures.

It would be possible to produce data of a similar nature by the use of a 20-sided die, in which one side is labelled F for failure and the other 19 S for success. This die would be thrown as many times as required, with the appropriate symbol being recorded according to the face that fell uppermost. The list of Ss and Fs so produced, while not being identical to the real data, would resemble it in a number of ways. Within limits allowed for by statistical variability, the proportion of failures would be similar, as would be the proportion of times that two failures appeared in succession. The lengths of the runs of successes between failures would be similarly distributed. It would be impossible to tell which data set was which from any statistical tests that could be applied. It is this fact that is used in Monte-Carlo modelling, or simulation as it is sometimes called.

In practice, dice are no longer used to generate the random data used in the simulation. Most calculators have a random number generator, often denoted RAN, and this can be used instead of a die. If the RAN button is pressed, a number between 0 and 1 will be displayed. If it is pressed a number of times in succession, different numbers will be displayed, apparently with no pattern. It is as if there is a die rolling facility in the calculator, that rolls a die, with many sides, and sends the results up to the display. In fact, the numbers are not truly random, but are calculated from a mathematical formula. They are called pseudo-random, and exhibit many properties of random numbers. All the numbers between 0 and 1 are equally likely, in the sense that if a large number of them were generated, and a histogram of the data drawn, it would be more or less flat. There is no significant correlation between each number and the next, etc. Designing good random number generators is a skilled task, and academic reputations have been made investigating the theory behind them.

Table 1.2 Simulation of two components in series

Simulation	Component 1		Component 2		System
	r	S or F	r	S or F	S or F
1	0.543	S	0.634	S	S
2	0.586	S	0.708	S	S
3	0.755	S	0.974	F	F
4	0.142	S	0.81	S	S
5	0.683	S	0.643	S	S
6	0.788	S	0.912	F	F
7	0.111	S	0.717	S	S
8	0.150	S	0.349	S	S
9	0.451	S	0.925	F	F
10	0.932	S	0.289	S	S
11	0.795	S	0.747	S	S
12	0.177	S	0.128	S	S
13	0.561	S	0.054	S	S
14	0.481	S	0.832	S	S
15	0.849	S	0.718	S	S
16	0.029	S	0.331	S	S
17	0.195	S	0.640	S	S
18	0.692	S	0.715	S	S
19	0.911	S	0.266	S	S
20	0.566	S	0.062	S	S

A random number generator can be used like the die discussed earlier. If the reliability of a component is R, then generate a random number r, and apply the following rule:

if

$$r < R \text{ then success}$$

$$\text{otherwise failure}$$

then the probability of success is exactly R, because all the numbers between 0 and 1 are equally likely. The use of the technique is best illustrated by examples.

Examples 1.1

1.1.1 Suppose that a system consists of two components in series, one with a reliability of 95 per cent, and the other with a reliability of 90 per cent.

A simulation, or simulated trial, consists of determining the success or failure of each component by the use of a die roll for each, and recording success for the system if both are a success and failure otherwise. Table 1.2 shows the results of twenty such trials. Just as with the

Table 1.3 Simulation of two components in redundancy

Simulation	Component 1		Component 2		System
	r	S or F	r	S or F	S or F
1	0.698	S	0.400	S	S
2	0.471	S	0.127	S	S
3	0.910	S	0.224	S	S
4	0.710	S	0.916	F	F
5	0.601	S	0.590	S	S
6	0.040	S	0.791	S	S
7	0.745	S	0.941	F	S
8	0.032	S	0.233	S	S
9	0.681	S	0.722	S	S
10	0.171	S	0.819	S	S
11	0.759	S	0.927	F	S
12	0.440	S	0.552	S	S
13	0.971	F	0.987	F	F
14	0.350	S	0.217	S	S
15	0.590	S	0.848	S	S
16	0.336	S	0.288	S	S
17	0.656	S	0.671	S	S
18	0.682	S	0.771	S	S
19	0.727	S	0.158	S	S
20	0.125	S	0.760	S	S

components, it is impossible to determine, by the use of statistical tests, that the sequence of Ss and Fs generated in this way is not real data, given that the random number generator is a good one (and most computers and calculators come with effective random number generators nowadays), and, more importantly, that the model is a good one.

1.1.2 Suppose the same two components are now in standby. Now there will only be a failure if both fail, and Table 1.3 shows the results of a simulation of such a system.

This brief introduction to Monte-Carlo modelling illustrates the method in the very simple situation in which a component either functions or fails. In practice, it may be necessary to obtain much more data about a system, such as its MTTF or MTBF, its availability, and the effect on these parameters of such things as MTTR, the number of repair teams, the spares holdings, etc. These aspects will be addressed in the following chapter on statistics. In that chapter the problems concerning the accuracy of the model and the source of data will also be discussed.

1.5.1 Perform a simulation on a system consisting of two components in series with reliabilities of 85 per cent and 92 per cent. Obtain ten data points, and compare your result with the analytic answer.

1.5.2 Perform Monte-Carlo simulations on the four-engined aircraft systems in Figs 15–17, assuming the engines are each 99 per cent reliable in each case. Compare your results with the analytic result.

1.5.3 Perform a Monte-Carlo simulation on the system shown in Fig. 1.19 (the bridge), assuming the following reliabilities for the components:

component	1	2	3	4	5
reliability (%)	95	90	85	80	75

These exercises can be done by a class or tutorial group together, the combined data of the students making a reasonably sized sample.

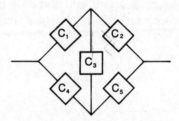

Fig. 1.19 The bridge.

Elementary statistics

2

2.1 Introduction

No text on reliability can be complete without some mention of statistics, and the good reliability engineer must be able to appreciate the meaning and implications of elementary statistical concepts such as mean, variance, distribution, etc. It is also essential that the results of others can be understood, and if necessary, criticized. This chapter introduces some elementary concepts of descriptive statistics, while some more advanced analyses and some of the ideas of statistical analysis and inference, such as significance and confidence, are left to later chapters.

2.2 Probability

2.2.1 Introduction

Probability is at the heart of any discussion on statistics. It is a concept that most people are familiar with, particularly when considered as chance, luck, likelihood, etc., or some other similar term. Having said that, when closely questioned, many people reveal that they are not really aware of the meaning of probability, or of its implications.

2.2.2 Definition

Consider the following (theoretical) experiment. A number of nominally identical items is tested, and the number of times the test is successful is recorded. The possible results for trials of different sizes are shown in Table 2.1.

It can be seen that the value in the right hand column, the proportion of successes, appears to be tending to some limit. This limit is the probability, giving the definition:

The probability of an event is the proportion of times that we observe that event in a large number of trials, or the proportion that we would expect to observe, were we able to observe a large number of trials.

Table 2.1 Results of trials of different sizes

Sample size	Number of successes
10	8
100	79
1000	881
10000	8764
100000	87561

(It is often impractical to actually carry out a large number of trials, or collect a large amount of data.)

Probability may also be expressed as a percentage. So, for example, it would appear that the probability of success in the trial of the example is between 0.87 and 0.88, or between 87 per cent and 88 per cent. Note that as probability is a proportion, it must lie between 0 and 1 inclusive.

So, for example, the author has found on questioning his students that about one student in twenty, on average, has had a catastrophic breakdown in his car in the previous year. (By catastrophic, it is meant one that stranded the driver by the side of the road, and necessitated an emergency call to a garage or one of the motoring organizations in order to be rescued.) This means that the average reliability of cars, in a year, is about 95 per cent, or that about one car in twenty will break down in a year.

Note that this is an average, and that not every class of twenty students will have exactly one person who has suffered a breakdown. Sometimes there will be no-one who has been that unlucky, and sometimes two or more. Later we will see just what the probability is of these, more complex, events happening.

It is necessary when considering complex situations to be able to combine probabilities, and the two basic laws concerning the way in which probabilities can be combined will be presented next.

2.2.3 The product rule

Suppose we have a simple situation in which two elements are in series. Call them C_1 and C_2, with reliabilities R_1 and R_2 respectively. Then the system of the two elements will only function if both of them are functioning. This can be written as:

The system functions *if* C_1 functions *and* C_2 functions and the probabilities (reliabilities) combine in the following way:

$$R_s = R_1 R_2$$

where R_s is the system reliability.

That is where this *and* that are needed, the probabilities are *multiplied* together. This is only true if the failures are statistically independent, that is, knowledge of the state of one of the components does not give information about the state of the other.

For example, the author's family has two cars. If one car will not start in the morning, then the other can be used. If both have a 1 per cent chance of not starting in the morning, what is the probability that neither will start in the morning?

Neither starting means that car number one will not start *and* car number two will not start, and so the answer is the product of the probabilities, i.e. 0.01×0.01, which is 0.0001, or 0.01 per cent.

This assumes that the failure at start of the two cars is independent. Now consider a second example, to illustrate the idea of lack of independence. After a thunder storm, it is observed that a light bulb has failed. It is believed that the probability of finding a failed bulb when the light is switched on is 1 in a thousand. This conclusion has been reached by collecting data for a year, and ignores any outside influences, such as the weather. Is the probability for the remaining light bulbs now still one in a thousand?

The answer is almost certainly, no. This is because thunder storms often overstress light bulbs. There is a higher probability that other bulbs have failed, as these failures are not independent, and there is an element of common cause failure. In practice, independence is assumed.

2.2.4 The summation rule

Consider the following situation. A switch can either fail closed or open. In a given period, it fails open 1 per cent of the time, and fails closed 2 per cent of the time. Then the total probability of failure is

$$1\% + 2\% = 3\%.$$

In this case, there are two events that can be observed, failed open or failed closed, and they are mutually exclusive (i.e. they cannot both happen). Then the probability that either of them is observed, i.e. this one *or* that one, is the sum of their probabilities. Notice that they must be mutually exclusive. The case where this is not so is dealt with below.

One consequence of this is the following, which can sometimes be used as a check on calculations. Suppose there are a number of mutually exclusive events that can be observed, and they are all that can happen:

$$E_1, E_2, E_3, \ldots$$

with probabilities:

$$P_1, P_2, P_3, \ldots$$

respectively. Then as this is all that can happen, and they are mutually exclusive, and as the total proportion must be exactly one, then

$$P_1 + P_2 + P_3 + \ldots = 1.$$

In the case that there are only two events of interest, and statisticians call these success and failure, with probabilities p and q respectively, this gives:

$$P(\text{Success}) + P(\text{Failure}) = 1$$

or

$$p = 1 - q. \tag{2.1}$$

Now consider the following situation. Suppose that a system with redundancy is made up of two components, C_1 and C_2, with reliabilities R_1 and R_2 respectively. Then the reliability of the two together is the probability that C_1 or C_2 is functioning, which according to the summation rule is

$$R_s = R_1 + R_2.$$

But clearly this could be greater than one, which is a nonsense. This apparent paradox arises because the functioning of the two components is not mutually exclusive, i.e. they may both be functioning at the same time (and may be expected to be doing so). This difficulty is overcome by considering the converse problem, that of calculating the failure probability. Let F, with the appropriate subscript, denote the probability of failure. Then,

$$F = 1 - R$$

and the system fails only if both components fail, that is to say, if C_1 fails *and* C_2 fails. Then by the product rule, this gives the failure probability of the two together as

$$F_s = F_1 \times F_2.$$

By substituting the values of R, and doing some elementary algebra, the reliability of the system can be obtained:

$$R_s = R_1 + R_2 - R_1 R_2.$$

This trick, of turning the problem round, and calculating the probability of failure rather than that of success, is very commonly used by statisticians in order to simplify the calculations.

Most situations can be dealt with by the appropriate use of the two formulae quoted above. More complex situations will require the use of both rules, and the application of some common sense and experience to describe the situation in terms of basic events whose probabilities satisfy the conditions necessary. Consider the following example to illustrate this point.

**Example
2.1** 2.1.1 A system consists of three components, C_1, C_2, and C_3, in 2/3 standby. Their reliabilities are 80 per cent, 85 per cent, and 90 per cent respectively. What is the reliability of the system?

   The system functions only if at least two, that is to say, two or more, of the three components are functioning. In order to make the point, a complete analysis of the system follows, but normally it would not be necessary to calculate both the probability that the system functions and the probability that it fails.

$$P(\text{All three are functioning}) = P(C_1 \text{ and } C_2 \text{ and } C_3 \text{ function})$$
$$= P(C_1 \text{ functions}) \times P(C_2 \text{ functions})$$
$$\times P(C_3 \text{ functions})$$
$$= 0.8 \times 0.85 \times 0.9$$
$$= 0.612$$

by the product rule.

$P(\text{Exactly two are functioning})$

$$= P((C_1 \text{ and } C_2 \text{ function and } C_3 \text{ fails})$$
$$\text{or } (C_1 \text{ and } C_3 \text{ function and } C_2 \text{ fails})$$
$$\text{or } (C_2 \text{ and } C_3 \text{ function and } C_1 \text{ fails}))$$
$$= P(C_1 \text{ and } C_2 \text{ function and } C_3 \text{ fails})$$
$$+ P(C_1 \text{ and } C_3 \text{ function and } C_2 \text{ fails})$$
$$+ P(C_2 \text{ and } C_3 \text{ function and } C_1 \text{ fails})$$

by the summation rule, as clearly the events are mutually exclusive,

$$= 0.8 \times 0.85 \times 0.1 + 0.8 \times 0.15 \times 0.9 + 0.2 \times 0.85 \times 0.9$$

by the product rule, and remembering that the probability of failure is one minus the probability of functioning,

$$= 0.086 + 0.108 + 0.153$$
$$= 0.329.$$

$P(\text{The system functions}) = P(\text{All three are functioning or exactly two are functioning})$

$$= P(\text{Three are functioning})$$
$$+ P(\text{Exactly two are functioning})$$

by the summation rule

$$= 0.612 + 0.329$$
$$= 0.941.$$

The problem could have been solved by considering the probability of failure, in which case

$$P(\text{The system fails}) = P(\text{All three fail } or \text{ exactly two fail})$$
$$= P(\text{All three fail}) + P(\text{Exactly two fail})$$

by the summation rule.

$$P(\text{All three fail}) = 0.003$$

by the product rule, and

$$P(\text{Exactly two fail}) = 0.056$$

by a combination of the product and summation rules, and so

$$P(\text{The system fails}) = 0.003 + 0.056$$
$$= 0.059.$$

Notice that

$$P(\text{The system functions}) + P(\text{The system fails}) = 0.941 + 0.059$$
$$= 1.0$$

as it should.

Exercises 2.1

2.1.1 A power system consists of a turbine, a generator, and a distribution board. Their reliabilities in a year are 90 per cent, 95 per cent, and 98 per cent respectively. If all of them are needed for the system to function, what is the reliability of the system?

2.1.2 A cooling system has three electrical pumps, with reliabilities in time T of 85 per cent, 90 per cent, 95 per cent. Determine the probabilities that:

(a) none of them fail;
(b) just one fails;
(c) exactly two fail;
(d) at most two fail;
(e) all of them fail.

2.1.3 A pumping system consists of three electric pumps and two electric generators to supply the electric power for the pumps. If just one pump and just one generator are required for the system to function, what is the system reliability if the reliability of the pumps and generators is 80 per cent and 90 per cent respectively?

2.3 Probability distributions

2.3.1 Discrete distributions

The idea of a distribution is more easily explained initially for discrete data, that is, when the data are obtained by counting, rather than by measuring something. Suppose we were to test a sample of 100 simple items, and count the number of failures we had. It may be that batches of the item have to be tested periodically, and this is done by testing a sample of 100 from each batch, and counting the number of failures. If the reliability of the item was 98 per cent, we would expect to see 2 failures in each sample, on average, but experience suggests that it would not be unusual to find just 1, or even none, from time to time, and if we were unlucky, to find 3 or 4 or even more. By collecting data over a suitable period, it would be possible to construct a histogram of the proportion of times that 0, 1, 2, etc. failures had been found in a batch of 100 such items. As proportions taken over a long period are probabilities, the values of the probabilities taken from the histogram gives the probability distribution of the number of failures in a sample of 100 items.

The distribution is simply presented as a list of all possible events, which must be exclusive, in this case the number of possible failures, and their probabilities. For a discrete case, this is the simplest explanation of a probability distribution, and can be presented in tabular form, as illustrated in Table 2.2.

Table 2.2 Frequency of observing numbers of failures in trials of 100 items

Number of failures	Frequency of occurrence
1	16
2	23
3	19
4	20
5	12
6	9
7	0
8	1
9 or more	0

Note that as the list includes all that we may observe, and the events are mutually exclusive, the sum of the probabilities is one, i.e.

$$\sum_i p_i = 1.$$

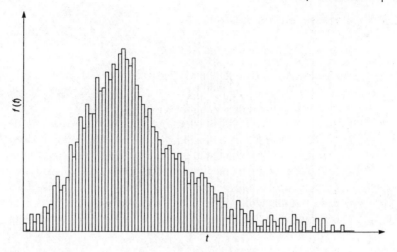

Fig. 2.1 Histogram of lifetimes.

2.3.2 Continuous distributions

For the continuous case, i.e. when the variable concerned is the result of measuring something, the situation is different. Consider the histogram of lifetimes of components that was presented in Fig. 1.1, in order to introduce the concepts of reliability. The collection of components that was tested in this situation is called a sample. The collection of all such components, possibly including those that have yet to be manufactured, is called a population, and the sample comes from a population. If a much larger sample were to be tested, the histogram would look similar, but instead of dividing the time axis into intervals of length one hour, and stating in which hour each component failed, we could divide the time axis into much smaller intervals. As a larger and larger sample were tested, the intervals could be made smaller and smaller, until the histogram looked like a continuous curve. This is illustrated in Figs. 2.1 and 2.2. The curve in this case gives the distribution of the lifetimes, and the histograms are approximations to it. It is denoted $f(t)$, and called the *probability density function*, or pdf.

It is the area under the curve which gives the probability, so that if $f(t)$ is the equation of the curve, then the probability of a component failing between times a and b is obtained by integrating between these times:

$$P(a \leqslant t \leqslant b) = \int_a^b f(t) \, dt.$$

This is shown as the shaded area in Fig. 2.3. As it is certain that the time

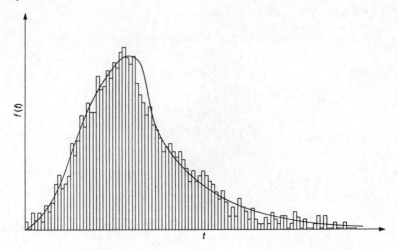

Fig. 2.2 Histogram of lifetimes and probability density function.

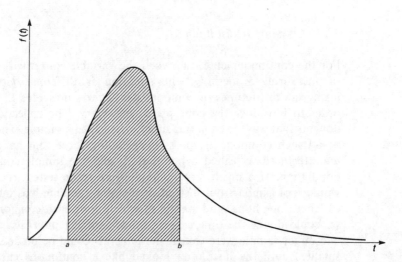

Fig. 2.3 Probability density function.

must be between zero and infinity, the total area under the curve must be one, i.e:

$$\int_0^\infty f(t)\,\mathrm{d}t = 1.$$

Notice that the probability of a component failing at time exactly t is zero.

2.3.3 Relationships between R, F, f, and λ

$F(t)$ is the probability that the component has failed by time t, or in other words that the lifetime or time to failure is less than t, and so

$$F(t) = \int_0^t f(\tau)\, d\tau.$$

This is called the cumulative function, or the ogive. Similarly, $R(t)$ is the probability that the item has not failed, i.e. that the lifetime or time to failure is greater than t, and so

$$R(t) = \int_t^\infty f(\tau)\, d\tau$$

$$= 1 - F(t)$$

and elementary calculus gives the inverse results:

$$f(t) = \frac{dF}{dt}$$

$$= -\frac{dR}{dt}.$$

The failure rate $\lambda(t)$ is defined by:

$$\lambda(t) = \frac{f(t)}{R(t)}$$

and has the property that the probability, P, that an item that is functioning at time t fails in the interval from t to $t + \delta t$ is given by

$$P = \lambda(t)\delta t.$$

Using the relationship between f and R, λ can be written

$$\lambda = \frac{-dR/dt}{R}$$

which is a differential equation for R if λ is known, the solution of which is

$$R = \exp - \left[\int_0^t \lambda(\tau)\, d\tau \right].$$

It will also be shown later that

$$\text{MTTF} = \int_0^\infty R(t)\, dt.$$

2.3.4 Distributions of maintenance times

If instead of considering times to or between failure, we wish to consider repair or maintenance times, if $m(t)$ is the pdf. of repair times, $M(t)$ is the maintainability, i.e. the probability that the item is restored in time t, $M(t)$ is the cumulative function,

$$M(t) = \int_0^t m(\tau)\, d\tau$$

and the repair rate μ is given by

$$\mu(t) = \frac{m(t)}{1 - M(t)}.$$

The repair or maintenance rate $\mu(t)$ has similar properties to the failure rate, λ, i.e. the probability, P, that an item that is in the failed state at time t is restored to the functioning state by time $t + \delta t$ is given by:

$$P = \mu(t)\, dt.$$

It is not generally found to be a useful concept, however, except when using Markov modelling.

Exercises 2.2

2.2.1 Consider the cooling system described in Question 2.1.2 of this chapter. If $n = 0, 1, 2,$ or 3 is the number of pumps functioning, write down the distribution of n in tabular form.

2.2.2 There are three electric lights in a room, each with a reliability of 95 per cent. Write down the distribution of the number of lights functioning. Check that the probabilities sum to one.

2.2.3 Repeat Question 2.2.2, but now with four lights, each with a reliability of 90 per cent.

2.2.4 The reliability of a certain system is given by

$$R = e^{-t^3}.$$

Write down the pdf of t and the failure rate (which is a function of t).

2.2.5 A system consists of two components in series, with constant failure rates λ_1 and λ_2. Write down

(a) the reliability of each component;
(b) the reliability of the system;
(c) the pdf of the time to failure of the system;
(d) the system failure rate;
(e) the MTTF.

2.2.6 Repeat Question 2.2.5, but this time assume the system consists of the same two components in active redundancy.

2.4 Summary measures

2.4.1 Introduction

It is useful when considering data and when considering distributions to consider values that summarize whatever is being considered. Generally it is sufficient to consider about where the data might be expected to lie, i.e. its position, and about how much noise there is in the data, i.e. variability. As a vehicle to illustrate the ideas that will follow, consider the data set given in Table 2.3, which comprises the data that gave rise to the histogram of Fig. 1.1.

Table 2.3 Life data

0.7	0.8	1.1	1.5	1.9	3.1	3.2	3.3	3.3	3.8
4.4	4.7	4.8	5.0	5.1	5.1	5.3	5.5	5.6	5.7
5.8	5.8	6.1	6.2	6.4	6.4	6.6	6.6	6.7	6.8
6.8	6.8	6.9	7.2	7.2	7.3	7.5	7.6	7.8	7.9
7.9	8.0	8.0	8.1	8.2	8.4	8.4	8.5	8.5	8.6
8.7	8.7	8.8	8.9	9.0	9.2	9.3	9.4	9.5	9.5
9.6	9.6	9.9	10.2	10.3	10.5	10.6	10.7	10.8	10.9
11.2	11.5	11.6	11.7	11.9	12.1	12.3	12.3	12.5	12.8
12.9	13.3	13.5	13.7	12.8	14.0	14.2	14.5	14.9	15.3
15.4	15.8	17.4	17.5	18.8	19.6	20.8	23.5	23.8	25.4

2.4.2 Measures of position

By a measure of position is meant a value that summarizes about where the data might be found, plus or minus some error term. The commonest one is the *arithmetic mean*, given by

$$\bar{t} = \frac{\sum t_i}{N}.$$

The data may be grouped, i.e. presented in the form shown in Table 2.4. The range of the data is divided into intervals of a suitable length, in this case one hour, and the numbers of data points falling in each interval, or frequency, are recorded. This is what is shown in Table 2.4.

There is a variety of conventions for dealing with data that corresponds exactly with the end points of an interval. In this case the interval is considered to include its left hand end point. Some authors will

Table 2.4 Life data, grouped

Interval†	Frequency	Interval†	Frequency
[0–1)	2	[13–14)	4
[1–2)	3	[14–15)	3
[2–3)	5	[15–16)	0
[3–4)	3	[16–17)	2
[4–5)	9	[17–18)	1
[5–6)	11	[18–19)	1
[6–7)	8	[19–20)	1
[7–8)	13	[20–21)	0
[8–9)	9	[21–22)	0
[9–10)	7	[22–23)	2
[10–11)	5	[23–24)	0
[11–12)	6	[24–25)	1
[12–13)	4		

† The notation [1–2) indicates that the interval includes its left hand end point but not its right hand one.

include the right-hand end point instead, while some will credit half a data point to each of the two intervals concerned.

For grouped data, the mean is given by

$$\bar{t} = \frac{\sum f_i t_i}{N}$$

where t_i is the mid point of the ith interval, f_i the frequency, or number of data in the ith interval, and N the sample size, given by

$$N = \sum f_i.$$

There are alternatives, such as the *median*, which is the 'middle value', i.e. the value of t such that half the values are less than and half more than that value. If there is an odd number of data in the sample, then the middle value is taken as the median, and if there is an even number of data, then the average of the two middle values is taken. If the data are grouped, then note that it is the area of each bar which is proportional to the number of readings in each group, not the height. Then the median is that value of t that halves the total area of the histogram. In some situations, the median is a better measure of position than the mean, particularly when the data are skewed.

A third alternative is the *mode*, which is the value that occurs most frequently. This can only be considered for grouped data.

For the data given above, the mean, median and modal values are given by, 8.5 hr, 8.7 hr, and 7.5 hr respectively.

In the example, a sample is being considered. When dealing with a population the pdf, $f(t)$, must be analysed.

For a distribution, the mean is called the *expected value* which is given by

$$E(t) = \int_0^\infty tf(t)\, dt$$

if the variable is continuous, and

$$E(i) = \sum ip_i$$

if the variable is discrete.

In the case of a continuous variable, it can be shown by using integration by parts that,

$$E(t) = \int_0^\infty 1 - F(t)\, dt$$

$$= \int_0^\infty R(t)\, dt.$$

For a population the expected value is often denoted μ (not to be confused with maintenance rate).

The median of the population is the solution for $\hat{t}$ of the equation

$$\int_0^{\hat{t}} f(t)\, dt = \int_{\hat{t}}^\infty f(t)\, dt = 0.5.$$

The mode is the value of t where $f(t)$ has a local maximum. Notice that there may be more than one mode, in which case the distribution is called multi-modal.

If $f(t)$ is reasonably well behaved, then the median lies between the mode and the mean.

If we were to take a sample from a population, then we would expect the sample mean to approximate to the expected value, and the sample median and mode to approximate to those of the population. We would also expect the approximation to be better for a larger sample. These ideas will be developed more fully in Chapter 5.

2.4.3 Measures of variability

It is necessary to know not only about where the data might lie, but also what kind of error term might be involved. When looking at a sample, the range might be taken as such a measure,

$$\text{Range} = t_{\max} - t_{\min}.$$

This has the disadvantage that it depends on only two readings, and the values of all the others do not even appear in the equation for the range.

Suppose we subtract the mean value from each t, and take the average of these, giving

$$\sum \frac{(t_i - \bar{t})}{N}.$$

It is not difficult to show that this is zero, as sometimes t_i is greater than the mean, and sometimes smaller, and so some of the values in the summation are positive, and some negative. In order to get rid of the differences in signs, the negative signs might just be thrown away, and the positive, numerical value of each term considered, giving

$$\frac{\sum |t_i - \bar{t}|}{N}$$

where $|x|$ means the positive, numerical value of x, obtained by throwing away the negative sign, if necessary, and is called the *modulus*, or *mod*, of x. This measure of variability is called the *mean deviation* of t. It is not difficult to calculate, particularly in the days before calculators and computers, but it is not frequently used, as it does not have many nice mathematical properties. It can also be used with the median instead of the mean, which in some ways is rather more consistent.

Rather the negative signs are removed by squaring each term (as the square of a negative number is positive), giving:

$$s^2 = \frac{\sum (t_i - \bar{t})^2}{N} \tag{2.2}$$

which is called the *sample variance*, and s is called the *sample standard deviation*. The sample variance is sometimes defined as

$$s^2 = \frac{\sum (t_i - \bar{t})^2}{N - 1}. \tag{2.3}$$

The reasons for this, and the exact use of the second formula, will be explained in Chapter 5.

For grouped data the variance is given by:

$$s^2 = \frac{\sum f_i (t_i - \bar{t})^2}{\sum f_i}.$$

For the sample of the example, the range is 23.7 hr, the mean deviation from the mean is 3.62 hr, and from the median is 3.54 hr, the standard deviation is 4.77 hr (using N in the denominator), and the standard deviation is 4.79 hr (using $N - 1$ in the denominator).

For a population, the *variance* is written σ^2, and is defined as

$$\sigma^2 = \int_0^\infty (t - \mu)^2 f(t) \, dt$$

for the continuous case, and for the discrete case,

$$\sigma^2 = \sum (i - \mu)^2 p_i.$$

For situations (samples and populations) that are not very skewed, it is often the case that about:

65 per cent of the data are within 1 standard deviation of the mean;

95 per cent of the data are within 2 standard deviations of the mean;

99 per cent of the data are within 3 standard deviations of the mean.

Exercises 2.3

2.3.1   For the life data shown in Table 2.5, find the mean, median, the mean deviations from the mean and median, variance, and the standard deviation.

Table 2.5 Life data (hr) from accelerated test trial

103	155	98	246	189	126	183	87	210	139

2.3.2   For the life data shown in Table 2.6, group the data, and find the summary measures as you did in Question 2.3.1. Also find the mode.

Table 2.6 Life data (10 000 revolutions) from trials of bearings

56	87	51	79	93	34	48	45	21	9
23	58	52	68	62	46	53	69	70	36
43	64	82	30	54	67	49	31	67	84
55	52	60	73	43	11	57	60	81	48
27	43	50	71	53	69	58	32	63	70

2.3.3   If the pdf of the life of an item is

$$f(t) = \lambda e^{-\lambda t}$$

find the MTTF and the reliability as a function of time using the formulae of the text.

2.3.4   Repeat question 2.3.3 when

$$f(t) = \lambda^2 t e^{-\lambda t}.$$

2.3.5   For a particular non-repairable system, it is believed that the failure rate is not constant, but varies exponentially with time, so that

$$\lambda = A e^{bt}$$

where A and b are constant. Find the relationship between reliability and time. If $A = 0.01$, and $b = 0.05$, and t is measured in days, what is the reliability after 10 days? If you initially had a large number of these systems, after how many days would you expect there still to be about half of them running?

2.5 Some common distributions used in reliability

2.5.1 Introduction

This section lists some of the commoner distributions that are met in reliability analysis. Where appropriate, some of their uses and properties are given. The first two distributions are of discrete variables, the remainder of continuous variables.

2.5.2 Binomial

Consider the following situation. It is decided to test a sample of n items, and the outcome of each test is either a success or a failure, i.e. each item either functions or fails to function. If the probability of success of each trial is p, then on average we would expect there to be np successes (this is how we defined p in Section 2.2.1). However, this is only an average figure, and if we are able to repeat the trial a large number of times experience tells us that sometimes there will be more than np successes and sometimes fewer. The binomial distribution gives the probability of there being exactly r successes for $r = 0, 1, 2, \ldots, n$, and states that if

$$P(\text{exactly } r \text{ successes}) = p_r$$

then

$$p_r = C_r^n p^r q^{n-r}$$

where

$$C_r^n = \frac{n!}{r!(n-r)!}$$

and

$$n! = n \times (n-1) \times (n-2) \times \ldots \times 3 \times 2 \times 1$$

and q is defined in eqn 2.1.

The expected value and variance are given by

$$E(r) = np$$

$$\text{variance}(r) = npq.$$

Examples 2.2

2.2.1 Consider the example mentioned at the start of 2.3.1. There a sample of 100 items was considered, each with reliability 98 per cent. We shall calculate the probability of obtaining 0, 1, 2, . . . failures in the trial, and as we are counting failures, a failure is considered a success in the above formula. (Funny people statisticians. The terms success and failure are technical terms in statistics, not to be confused with the emotive ones used in everyday life.) So

$$n = 100, \quad p = 0.02, \quad q = 0.98.$$

The expected (or average) number of failures is given by

$$np = 2$$

and the variance is

$$npq = 1.96$$

and so the standard deviation is

$$\sqrt{npq} = 1.4$$

and:

$$p_0 = 0.98^{100} \qquad\qquad\qquad\qquad = 0.133 \ (0! = 1 \text{ and } x^0 = 1);$$

$$p_1 = 100 \times 0.98^{99} \times 0.02 \qquad\qquad = 0.271;$$

$$p_2 = \frac{100 \times 99}{2} \times 0.98^{98} \times 0.02^2 \qquad = 0.273;$$

$$p_3 = \frac{100 \times 99 \times 98}{2 \times 3} \times 0.98^{97} \times 0.02^3 \ = 0.182;$$

etc.

2.2.2 Consider the four engined aircraft mentioned in Chapter 1. Suppose that it can continue its mission if it has at least two engines still functioning. If the reliability of each engine is 99 per cent, what is the probability of the aircraft being able to complete its mission?

It can complete its mission if there are two, or three, or all four engines still functioning. In this case, success is an engine running, and

$$n = 4, \quad p = 0.99, \quad q = 0.01.$$

Then

$$p_4 = 0.99^4 \qquad\qquad\qquad = 0.960\,596;$$

$$p_3 = 4 \times 0.99^3 \times 0.01 \qquad = 0.038\,812;$$

$$p_2 = 6 \times 0.99^2 \times 0.01^2 \quad = 0.000\,588.$$

The probability of two or more engines functioning is the sum of these by the summation formula, so

$$P(2 \text{ or more engines function}) = 0.999\,996.$$

In practice, it would have been easier in this case to calculate the unreliability of the system (i.e. the probability of failure) as the probability of there being one or no engines functioning. This is left as an exercise for the reader, who can check that it is $1 - 0.999\,996 = 4 \times 10^{-6}$.

The use of the binomial has been well illustrated by the examples. Calculating the probabilities of the numbers of defectives in samples, or dealing with k-out-of-n redundancy is the commonest use that the reliability engineer will see.

2.5.3 Poisson

When counting a number of incidents, such as failures, that occur independently in a fixed time period, the Poisson distribution is used. So it may be the number of failures to a fleet of vehicles in a week, or the number of accidents on a fixed stretch of road in a month, etc. There is no prior maximum number of such incidents that may be observed, as there is with the binomial distribution, hence r is unbounded. The average number of incidents we would expect is m.

In this case,

$$p_r = \mathrm{e}^{-m}\frac{m^r}{r!}, \quad r = 1, 2, 3, \ldots$$

The expected value and variance are given by

$$E(r) = m;$$

$$\text{variance}(r) = m.$$

The independence criterion means that the likelihood of an incident in the near future is not dependent on the occurrence or non-occurrence of one in the recent past. This is the case in the two examples mentioned above, as the presence or absence of failures in one part of a fleet has no effect on the probability of failures in another part, and similarly for accidents on a road.

The Poisson is a good approximation to the binomial when n is large and p is small, in which case the expected values are made identical, i.e. m is put equal to np.

Examples 2.3

2.3.1 Over a year, the number of failures of a fleet of vehicles each month was:

J	F	M	A	M	J	J	A	S	O	N	D
1	2	1	0	3	1	0	3	2	2	1	2

What is the probability of exactly 0, 1, 2, ... failures in a given month?
The average number of failures per month is 1.5, so using the Poisson
distribution with $m = 1.5$ gives:

$$p_0 = e^{-1.5} \qquad\qquad = 0.223;$$

$$p_1 = e^{-1.5} \times 1.5 \qquad = 0.335;$$

$$p_2 = e^{-1.5} \times \frac{1.5^2}{2} \qquad = 0.251;$$

$$p_3 = e^{-1.5} \times \frac{1.5^3}{2 \times 3} \qquad = 0.126;$$

etc.

2.3.2 Use the Poisson approximation to the binomial to calculate the
probabilities of 0, 1, 2 ... failures in Example 2.2.1, and compare the
results with those obtained using the binomial distribution.
Using the Poisson distribution, $m = np = 2$

Poisson		binomial
$p_0 = e^{-2}$	$= 0.135$	0.133
$p_1 = e^{-2} \times 2$	$= 0.271$	0.271
$p_2 = e^{-2} \times \dfrac{2^2}{2}$	$= 0.271$	0.273
$p_3 = e^{-2} \times \dfrac{2^3}{2 \times 3}$	$= 0.180$	0.182

The Poisson is used in the analysis of incidents such as breakdowns,
failures, accidents, etc. when the incidents are independent. The
examples illustrate this application. It can also be used to analyse the
number of flaws in a sample of material (a length of pipe, for example)
if it is believed that the flaws occur independently.

2.5.4 Exponential

The pdf is given by

$$f(t) = \lambda e^{-\lambda t}, \quad 0 \leqslant t < \infty$$

$$E(t) = \frac{1}{\lambda}$$

$$\text{variance} = 1/\lambda^2$$

the failure rate $= \lambda$ (constant)

$$R(t) = e^{-\lambda t}.$$

The exponential distribution is commonly used to describe times to or between failures, in which case $E(t)$ is the MTTF or the MTBF. The constant failure rate model assumes the failures are independent, and so make the same assumptions as the Poisson distribution. If a repairable system has a constant failure rate, the times between failure are exponentially distributed, and the number of failures in a time period T is Poisson distributed with mean λT.

Example 2.4

2.4.1 A system is believed to have a constant failure rate, and a MTBF of 1000 hr. Then

$$f(t) = 0.001e^{-0.001t}$$

and the reliability is given by

$$R(t) = e^{-0.001t}$$

so that at times $t = 10, 100$, and 500 hr the reliabilities are 99 per cent, 90.5 per cent, and 60.1 per cent.

The use of the constant failure rate model has already been discussed in Chapter 1.

2.5.5 Gamma

$$f(t) = \frac{\lambda^{\alpha}}{\Gamma(\alpha)} t^{\alpha-1} e^{-\lambda t}, \quad 0 \leqslant t < \infty$$

where

$$\Gamma(\alpha) = \int_0^{\infty} t^{\alpha-1} e^{-t} \, dt$$

and is called Gamma of α.

$$E(t) = \frac{\alpha}{\lambda};$$

$$\text{variance}(t) = \alpha/\lambda^2.$$

The expression for $R(t)$ cannot be written down in closed form, except in the special case that α is an integer (a whole number), when it is a long winded algebraic expression. Similarly for the failure rate, but the failure rate is increasing with time if $\alpha > 1$ and decreasing if $\alpha < 1$.

In the case that α is an integer, the distribution is known as the Erlang distribution. In this case, it is the distribution of a time that is made up of the sum of identically, exponentially distributed times. Such a situation arises when n items are in dormant redundancy, and only one of them is needed at any one time. If they are identical, and have the same constant failure rate, with negligible probability of failure in the dormant state, then the time to failure of the system composed of these items is Erlang distributed, with α taking the value n, as the lifetime of the system is the sum of the lifetimes of the n components.

In the special case that α is an integral multiple of $1/2$, and λ is a half, the distribution is called the χ^2 distribution (pronounced chi-squared). So if x is χ^2 distributed, then

$$f(x) = \frac{1}{2^{n/2}\Gamma(n/2)} x^{(n/2)-1} e^{-x/2}.$$

The parameter n is called the number of *degrees of freedom*, and is sometimes written v. It is impossible to write down an analytic expression for the cumulative distribution in this case unless n is an even number, and the expression is complex unless n is small. If the χ^2 distribution is required, it is tabulated, and the user must refer to the tables.

Example 2.5 2.5.1 Suppose two identical systems with failure rate λ are in standby redundancy. Then the pdf of the lifetime of the system assuming no switching or dormant failure is given by

$$f(t) = \lambda^2 t e^{-\lambda t}$$

and the reliability is (integrating by parts)

$$R(t) = [1 + \lambda t] e^{-\lambda t}.$$

This use of the gamma distribution, while useful, is not very common. Rather, the χ^2 distribution is the commonest application of the gamma, and that is for statistical data analysis. This will be dealt with in Chapter 5. The gamma is also useful when doing Bayesian analysis, and this is explained in Chapter 6.

2.5.6 Beta

The variable x is said to have a beta distribution if its pdf is given by

$$f(x) = \frac{\Gamma(\alpha + \beta)}{\Gamma(\alpha)\Gamma(\beta)} x^{\alpha-1} (1-x)^{b-1}, \quad 0 \leqslant x \leqslant 1.$$

The expected value and variance are given by

$$E(x) = \frac{\alpha}{\alpha + \beta};$$

$$\sigma^2 = \frac{\alpha\beta}{(\alpha + \beta)^2 (1 + \alpha + \beta)}.$$

The integrals needed to calculate the cumulative function cannot be done analytically except in special cases, and like the χ^2 it is tabulated. An illustration of its use will not be given here, but it is used in Bayesian analysis, and its use is shown in the Section 6.3.

2.5.7 Weibull

The two parameter Weibull is given by

$$f(t) = \frac{\beta}{\eta}\left(\frac{t}{\eta}\right)^{\beta-1} \exp\left(-\left(\frac{t}{\eta}\right)^{\beta}\right), \quad 0 \leqslant t < \infty;$$

$$E(t) = \eta\Gamma\left(1 + \frac{1}{\beta}\right);$$

$$\mathrm{var}(t) = \eta^2\left[\Gamma\left(1 + \frac{2}{\beta}\right) - \Gamma\left(1 + \frac{1}{\beta}\right)^2\right];$$

$$R(t) = \exp\left(-\left(\frac{t}{\eta}\right)^{\beta}\right);$$

and

$$\lambda(t) = \frac{\beta}{\eta}\left(\frac{t}{\eta}\right)^{\beta-1}.$$

It is sometimes generalized to the three parameter Weibull distribution, in which case

$$f(t) = \frac{\beta}{\eta}\left(\frac{t - t_0}{\eta}\right)^{\beta-1} \exp\left(-\left(\frac{t - t_0}{\eta}\right)^{\beta}\right) t_0 \leqslant t;$$

$$E(t) = t_0 + \eta\Gamma\left(1 + \frac{1}{\beta}\right);$$

$$\mathrm{var}(t) = \eta^2\left[\Gamma\left(1 + \frac{2}{\beta}\right) - \Gamma\left(1 + \frac{1}{\beta}\right)^2\right];$$

$$R(t) = \exp\left(-\left(\frac{t - t_0}{\eta}\right)^{\beta}\right);$$

and

$$\lambda(t) = \frac{\beta}{\eta}\left(\frac{t - t_0}{\eta}\right)^{\beta-1}.$$

The Weibull distribution is very commonly used in reliability analysis, and is very useful to describe increasing and decreasing failure rate, so much so that Section 4.4 is devoted to a discussion of it and how the parameters can be estimated from data. Further discussion is left to Section 4.4.

2.5.8 Normal

The pdf is:

$$f(t) = \frac{1}{\sigma\sqrt{2\pi}}\exp\left(-\frac{1}{2}\left(\frac{t-\mu}{\sigma}\right)^2\right), \quad -\infty < t < \infty;$$

$$E(t) = \mu;$$

$$\text{variance}(t) = \sigma^2.$$

The reliability and failure rate cannot be written in closed form, and tables or a computer or calculator package are needed to obtain values of the reliability if the normal distribution is used. It is not popular with reliability analysts, as t can take negative values, but it is sometimes used, in which case such values are ignored. i.e. it is truncated at $t = 0$. This does effect the values of the expected value and variance. Transformations of times are sometimes normally distributed, for example see below.

The normal distribution is more commonly used when analysing such parameters as the strengths of material, wear, the diameters of shafts, and other dimensions, when it is extremely useful. In these cases the expected value is usually very large compared with the variance, and the probability of a negative value is very small indeed. There can be very good theoretical reasons for using the normal distribution. See Carter (1986) for a more detailed discussion on the use of the normal distribution in reliability and engineering.

No worked examples are offered here, as it will not be used in the book, but it is here for completeness, particularly in view of the comments on the log-normal distribution and maintenance times. The Weibull distribution with shape parameter 3.44 approximates very closely to the normal distribution.

2.5.9 Log-normal

The variable t is said to be log-normally distributed if the natural logarithm of t, $\ln(t)$, is normally distributed. In this case,

$$f(t) = \frac{1}{t\sigma\sqrt{2\pi}}\exp\left(-\frac{1}{2}\left(\frac{\mu-\ln(t)}{\sigma}\right)^2\right), \quad 0 \leqslant t < \infty;$$

$$E(t) = \exp(\mu + \sigma^2/2);$$

$$\text{variance}(t) = \exp(2\mu + \sigma^2)(\exp(\sigma^2) - 1).$$

The log-normal distribution is frequently used when analysing repair or maintenance times. The cumulative function and rate function cannot be written down in closed form.

The parameter μ is the expected value of $\ln(t)$, and as it is also the median, it must transform to the median of t. This gives e^μ as the median for t.

The mean of the natural log of the sample values is approximately equal to μ. This gives

$$\mu \approx \sum \ln(t)/n$$

for a sample of size n. This is equal to

$$\ln(\Pi t)/n$$

where Π is the product, and by the fundamental property of logs,

$$\ln((\Pi t)^{1/n})$$

again by the property of logs. The term inside the log, $(\Pi t)^{1/n}$ is called the *geometric mean* (as opposed to the arithmetic mean) of the sample, and for this distribution, is an estimate of the median. When analysing maintenance times, the geometric mean (as well as the arithmetic mean) is frequently quoted in the results. The parameter σ is called the dispersion of the distribution.

2.5.10 Student's t

The t distribution, as it is sometimes known, is a distribution that is derived theoretically, and data is rarely, if ever, t distributed. For that reason the pdf and other parameters are not given here. The interested student can consult DeGroot (1986), for example, or any other good statistics text. The t distribution is, however, used in data analysis under certain circumstances, and is included here because it is used in Section 5.7. Like so many distributions, it is tabulated rather than calculated each time it is used.

2.5.11 Uniform

The uniform distribution is the one in which all values are equally likely over the finite interval of interest. Then

$$f(t) = 1/(b - a), \quad a \leqslant t \leqslant b;$$
$$R(t) = (b - t)/(b - a);$$
$$E(t) = (b + a)/2;$$
$$\text{variance}(t) = (b - a)^2/12.$$

The uniform distribution is rarely used to analyse data directly, but is useful in theoretical analysis. In particular, it is used when performing Monte-Carlo simulation.

For a number of the distributions mentioned above, it is difficult, if not impossible, to do the integrals necessary by analytic means. Many textbooks contain tables of these distributions, or tables can be obtained commercially, or many computer packages will give the values concerned.

Exercises 2.4

2.4.1 A sample of 80 flares is taken from a large batch which is believed to have a reliability of 97 per cent. What is the probability of more than three duds in the sample?

2.4.2 A power generation system in a factory has six identical generators, each with a reliability of 90 per cent. For some purposes only four are needed. What is the probability of four or more of the generators functioning?

2.4.3 The average number of serious malfunctions per month on an oilrig is 1.8. What is the probability of more than four such malfunctions in any given month?

2.4.4 A component has a constant failure rate with an MTTF of 3500 hr in a given application. What is the mean number of failures per year, if a failed component is replaced immediately it fails (there are 8766 hr in a year)? Use the Poison distribution to calculate the probability of more than four failures in any given year.

2.4.5 A round the world yachtsman knows that one of the components in his steering gear has a constant failure rate and a mean distance to failure of 8000 miles. Taking the distance he expects to travel as 25 000 miles, what is the probability of the component in question causing a total breakdown of the steering gear if he sets out with four spares, as well as the one in use?

2.4.6 A component has a constant failure rate of .002 failures per hour. What is the probability of it failing between 50 hr and 100 hr?

2.4.7 A system consists of two identical components in standby, with constant failure rate and MTTF of 1000 hr. What is the probability of it failing between 100 hr and 200 hr?

2.4.8 An electric fuse must fail (i.e. melt) when the current is between 12.5 and 13.4 amps. It is known that the current that causes a particular fuse design to fail is three parameter Weibull with minimum current of 11 amps, shape parameter 4.2 and scale parameter 2.1 amps. Calculate the reliability of the fuse.

2.6 Simulating time in Monte-Carlo modelling

2.6.1 Introduction

In Section 1.9 the idea of Monte-Carlo modelling was introduced, and illustrated by some examples in which only the success or failure of the components and systems were considered. In this section it is shown how time can be simulated, so that time related parameters, such as MTTF, MTBF, availability etc., can be estimated from the model.

Just what is meant by simulating time is the following. Suppose we have a data set of lifetimes, obtained from a trial. The set can be used to draw a histogram like that in Fig. 1.1. Then a simulation of the component would use the random number generator in a calculator or computer to produce a second data set that would be similar to the first. Although not exactly the same, the means, variances, and other summary measures would be close, and the histograms would look similar. It would be impossible to tell by any statistical test which was the real data and which the simulated, just as with the situation described in Section 1.9.

2.6.2 Simulating a distribution

It can be shown that if the distribution of the parameter of interest is known, or assumed, then if $f(t)$ is the pdf and $R(t)$ is the reliability, then if r comes from a random number generator, so that r is uniformly distributed, then the parameter t which is the solution of the equation:

$$R(t) = r \qquad (2.4)$$

has a distribution with pdf $f(t)$.

Examples 2.6 2.6.1 Suppose two components C_1 and C_2 are in series, that both components have constant failure rate with MTTF of 100 hr, and that there is no repair. It is required to simulate the system in order to estimate its MTTF.

A single simulation of the system is as follows. Two random numbers, r_1 and r_2 are generated, one for each component. The equations:

$$e^{-\lambda t_i} = r_i$$

must be solved in order to generate the simulated lifetimes t_i (because the components have a constant failure rate, the reliability is given by $e^{-\lambda t}$), for $i = 1, 2$, and $\lambda = 0.01$. As the components are in series, the minimum of these two values gives the system life for that simulation. In this case the equations are easy to solve algebraically, and give:

Table 2.7 Simulation of two components in series

Simulation	Component 1		Component 2		System
	r	Life	r	Life	Life
1	0.815	20	0.155	185	20
2	0.275	129	0.548	60	60
3	0.179	171	0.197	162	162
4	0.322	113	0.807	21	21
5	0.280	127	0.188	167	127
6	0.726	32	0.621	48	32
7	0.765	27	0.287	125	27
8	0.826	19	0.107	223	19
9	0.907	10	0.323	112	10
10	0.130	204	0.046	307	204

$$t_i = -\frac{\ln(r_i)}{\lambda}$$

$$= -\text{MTTF} \ln(r_i)$$

Ten simulations are shown in Table 2.7.

2.6.2 Suppose that two components C_1 and C_2 are in redundancy. Their lifetimes are Weibull distributed with scale parameters 15 and 20 hr and shape parameters 2.5 and 3.5 respectively. In order to generate the times, it is necessary to solve the equation

$$\exp -\left(\frac{t_i}{\eta_i}\right)^{\beta_i} = r_i$$

which can be solved algebraically, to give

$$t_i = -\eta_i [\ln(r_i)]^{1/\beta_i}.$$

Two sets of ten simulations of each situation are shown in Table 2.8, one each for the situation hot and cold redundancy. For the former, the maximum of the two simulated times is taken, while for the latter, the sum of the times is taken, for the time to failure of the system. In this case it is assumed that there is no dormant failure, and no switching failure. These can be simulated, as is shown in the next example.

2.6.3 Consider Example 2.6.2, assuming that C_2 is in dormant standby, with a MTTF of 50 hr, and that the failure rate in the dormant state is constant. Suppose further that there is a probability of a switching failure of 5 per cent. Table 2.9 shows ten simulations of the system.

When C_1 fails, at a time shown in Column 3, C_2 only comes on line if there is no switching failure shown in Columns 6 and 7. If C_2 does not come on line, the system time to failure is the time to failure of

Table 2.8 Simulations of two components in hot and cold redundancy

Simulation	Component 1		Component 2		System	
	r	Life	r	Life	Hot	Cold
1	0.677	10	0.333	21	21	31
2	0.569	12	0.721	14	14	26
3	0.081	22	0.480	18	22	40
4	0.503	13	0.091	26	26	39
5	0.955	4	0.788	13	13	17
6	0.162	19	0.317	21	21	40
7	0.027	25	0.367	20	25	45
8	0.357	15	0.929	9	15	24
9	0.038	24	0.543	17	24	41
10	0.136	20	0.820	13	20	33

Table 2.9 Simulation of cold redundancy with possibility of switching failure

Simulation	Component 1		Component 2		Switch		System
	r	Life	r	Life	r	S or F	
1	0.369	15	0.431	19	0.10	S	34
2	0.456	14	0.370	20	0.15	S	34
3	0.146	19	0.800	13	0.64	S	32
4	0.187	18	0.535	17	0.75	S	35
5	0.878	7	0.940	9	0.47	S	16
6	0.452	14	0.763	14	0.66	S	28
7	0.116	20	0.071	26	0.96	F	20
8	0.597	12	0.056	27	0.90	S	39
9	0.636	11	0.224	22	0.40	S	33
10	0.874	7	0.611	16	0.14	S	23

C_1, otherwise, if C_3 does come on line, a time to failure is generated for it, Column 5, and the time to failure of the system is the sum of the times in Columns 1 and 5, shown in Column 8.

This is not the only way in which times can be simulated, and in some situations the particular properties of the distributions are used when simulating times drawn from them. The means of simulating times from the distributions discussed above are presented in Section 2.6.4, although the method described in this section will always work, if eqn 2.4 can be solved.

2.6.3 Repairable systems and other possible model inputs

It is, of course, possible to simulate repairable systems, in which case a number of factors concerning the model must be determined in advance.

The analyst must know the time to repair, which can be deterministic or random. If random, the time to repair can be determined in a similar fashion to the time to failure, as described in the examples above.

The problems involved with spares holdings can be studied, or may be ignored (i.e. assume there is always a spare available when one is needed). Otherwise, account may be taken of the spares available, either by keeping a count of the spares held or by determining the spares availability by random means.

The number of repair teams available and hence the possible need to queue for repair can be put into the model. Here again, as with the spares availability problem, it is possible either to keep a count of the teams available and what they are doing, or the repair team availability can be determined in a statistical manner. In either case, the queue length and the time spent waiting for repair may well be a parameter of interest. If there are a limited number of repair teams, then repair priorities can also be modelled.

When a system is down because one or more of its subsystems are failed, it may or may not be the case that the remaining systems are kept running. Either situation can be simulated.

It is possible to take account of added stresses on surviving elements when a system is partially failed, usually by shortening remaining lifetime by a predetermined factor, although if the data were available, it would be possible to develop a more subtle model.

Switching failure and dormant failure have already been mentioned in Example 2.6.3.

Example 2.7

2.7.1 Consider a generator system consisting of a petrol engine and two generators in standby. The RBD of the system is shown in Fig. 2.4. All the subsystems have a constant failure rate, the petrol engine has an MTTF of 100 hr and the generators an MTTF of 150 hr when they are both running. If one has failed, the life of the survivor is shortened by 20 per cent until the failed item is brought back on line. (Another way of putting it would be to consider that each generator has a certain amount of life and that during any period in which only one is running, the other uses up its life at a faster rate.) If both generators have failed, the engine is not switched off, but keeps running, and hence can possibly fail. If the engine has failed, however, the generators are not used, and so cannot fail, and during the time the engine is being repaired, the generators are not using any life. There is only one repair team, and the engine is always repaired in preference to the generators. Repair

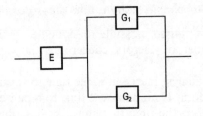

Fig. 2.4 Residual block diagram of generator system.

Table 2.10 Simulation of system involving repair

Number	Description	Time (hr)	Comments
1	E fails	3	Generates event 4. Events 2 & 3 become events 5 & 6 respectively.
2	G_1 fails	11	Delayed to event 5
3	G_2 fails	73	Delayed to event 6
4	E back up	8	New lifetime of 117 hr generates event 7 at 125 hr.
5	G_1 fails	16	Repair time of 2 hr generates event 8. During repair the life of G_2 is shortened by 1 hr (after rounding) to 77 hr (event 9). Event 7 becomes event 10.
6	G_2 fails	78	Becomes event 9
7	E fails	125	Becomes event 10
8	G_1 back up	18	Generate new life for G_1 of 469 hr, and hence event 11 at 487 hr.
9	G_2 fails	77	Generate event 12 at 79 hr. Events 10 & 11 becomes events 13 & 14 respectively.
10	E fails	125	Becomes event 13
11	G_1 fails	450	Becomes event 14
12	G_2 back up	79	New life of 150 hr, and so event 15 at 229 hr. Event 14 becomes event 16.
13	E fails	125	Generates event 17 at 127 hr. Events 15 & 16 become events 18 & 19 respectively.
14	G_1 fails	450	Becomes event 16
15	G_2 fails	229	Becomes event 18
16	G_1 fails	450	Becomes event 19
17	E back up	127	New life of 108 hr generates event 20 at 235 hr. Event 19 becomes event 21.
18	G_2 fails	234	New event 22 at 236 hr. Event 21 becomes event 23.
19	G_1 fails	450	Becomes event 21
20	E fails	235	Generates event 24 at 240 hr. Event 22 postponed to event 25 at 241 hr, and event 23 postponed to event 26 at 455 hr.
	etc.		

times are deterministic, and are 5 hr for the engine and 2 hr for a generator.

The results of the simulation are shown in Table 2.10. The technique used here is known as event stepping, where the times generated, either deterministically or statistically, generate events, which in this case are either a system failing or being repaired. Events are dealt with in the order in which they happen in the simulation, and each event generates a further event, and event times are modified, usually delayed, and appear further down the list, as a result of having to queue for repair etc. when higher priority items fail.

Exercises 2.5

2.5.1 Simulate the system in Example 2.7.1, but with repair. Assume it takes 1 hr to do the repair, that there are two repair teams, and that a failed system continues to operate.

2.5.2 Simulate the bridge shown in Fig 1.19. Assume the components have constant failure rates with the following MTTFs.

Component	1	2	3	4	5
MTTF (hr)	200	150	120	100	75

Combine the class results, if possible, to obtain an MTTF for the system.

2.6.4 Simulating other distributions

In this section the means of simulating the distributions of Section 2.5 are presented.

Binomial For the binomial, it is an easy matter to simulate each trial,

Table 2.11 Simulated binomial variables

Trial	Simulation											
	1		2		3		4		5		6	
	r	S/F	r	S/F	r	S/F	r	S/F	r	S/F	r	S/F
1	0.05	S	0.06	S	0.54	S	0.15	S	0.56	S	0.53	S
2	0.02	S	0.93	F	0.67	S	0.40	S	0.27	S	0.89	F
3	0.84	F	0.79	S	0.48	S	0.36	S	0.68	S	0.75	S
4	0.34	S	0.76	S	0.89	F	0.51	S	0.69	S	0.96	F
5	0.15	S	0.12	S	0.80	F	0.61	S	0.81	F	0.15	S
6	0.61	S	0.23	S	0.97	F	0.28	S	0.75	S	0.66	S
7	0.38	S	0.24	S	0.19	S	0.20	S	0.75	S	0.31	S
8	0.19	S	0.71	S	0.99	F	0.04	S	0.47	S	0.31	S
9	0.08	S	0.85	F	0.54	S	0.75	S	0.29	S	0.97	F
10	0.68	S	0.41	S	0.95	F	0.08	S	0.42	S	0.43	S
Number of failures	1		2		5		0		1		3	

with appropriate probability of success, and repeat the simulation the correct number of times. Table 2.11 shows six simulations of the binomial for $p = 0.2$ and $n = 10$.

Poisson If the number of events in a given time is Poisson, then the time between the events is exponential. So if it is required to generate a variable that is Poisson distributed with mean m, this can be modelled by means of an exponential variable with unit failure rate and time period m. Continue to simulate exponentials, until the cumulative sum exceeds m, recording the number of simulations, n, necessary to do this, in which case $n - 1$ is the variable required. This is illustrated in Table 2.12 for the case $m = 3$.

Table 2.12 Simulated Poisson variables

Simulation	r	Exponential	Cumulative	n	Poisson
1	0.768	0.264	0.264		
	0.991	0.009	0.273		
	0.928	0.075	0.348		
	0.085	2.467	2.816		
	0.427	0.850	3.665	5	4
2	0.156	1.923	1.923		
	0.348	1.056	2.979		
	0.846	0.168	3.147	3	2
3	0.092	2.381	2.381		
	0.824	0.193	2.574		
	0.053	2.940	5.514	3	2
4	0.989	0.011	0.011		
	0.071	2.638	2.650		
	0.334	1.096	3.746	3	2
5	0.581	0.543	0.543		
	0.958	0.043	0.586		
	0.531	0.632	1.218		
	0.721	0.328	1.546		
	0.078	2.553	4.099	5	4
6	0.105	2.251	2.251		
	0.166	1.793	4.044	2	1
7	0.231	1.467	1.467		
	0.790	0.235	1.702		
	0.823	0.194	1.896		
	0.181	1.710	3.606	4	3
8	0.044	3.124	3.124	1	0
9	0.156	1.857	1.857		
	0.072	2.620	4.477	2	1
10	0.262	1.340	1.340		
	0.839	0.176	1.515		
	0.161	1.826	3.341	3	2

Table 2.13 Simulated Erlang variables

Simulation	Exponentials						Erlang
	r	Time	r	Time	r	Time	Total
1	0.171	176	0.803	22	0.991	1	199
2	0.927	7	0.670	40	0.298	121	168
3	0.266	132	0.877	13	0.794	23	169
4	0.825	19	0.479	74	0.587	53	146
5	0.616	48	0.085	246	0.250	139	433
6	0.840	17	0.886	12	0.922	8	37
7	0.010	458	0.928	7	0.524	64	530
8	0.341	108	0.623	47	0.097	233	388
9	0.476	74	0.128	206	0.381	96	376
10	0.213	154	0.039	323	0.857	15	494

The exponential has already been dealt with.

Gamma The gamma distribution when α is a fraction is difficult to deal with, but is rarely needed. The Erlang, i.e. when α is an integer can be dealt with easily. Because of the property discussed in Section 2.4.5, that the Erlang distribution with shape parameter α and scale parameter λ is the sum of α independent exponentials with failure rate λ, generate α exponentials with failure rate λ, and sum them. This is illustrated if Table 2.13, with $\lambda = 0.01$ and $n = 3$.

The Weibull distribution has already been dealt with.

Normal A well known property of the normal distribution is that if a number of random variables are added together, then the result is approximately normal, with mean the sum of the means and variance the sum of the variances. (This is known as the central limit theorem in statistics.)

The uniform distribution between 0 and 1 has mean $\frac{1}{2}$ and variance $1/12$, so the sum of twelve uniform variables is approximately normal with mean 6 and variance 1. The approximation is very good, and quite suitable for most purposes. In order to simulate a normally distributed variable with mean μ and standard deviation σ, carry out the following steps:

1. Generate 12 uniformly distributed variables, and sum them (the random number generators on calculators and computers generate numbers uniformly distributed between 0 and 1).

2. Subtract 6, which makes the mean 0.

3. Multiply by σ, which changes the standard deviation to σ and leaves the mean unchanged.

Table 2.14 Simulated normal and log-normal variables

Simulation	1	2	3	4	5	6	7	8	9	10
1	0.08	0.79	0.07	0.80	0.33	0.24	0.83	0.76	0.51	0.27
2	0.34	0.38	0.38	0.77	0.79	0.21	0.86	0.50	0.78	0.12
3	0.70	0.60	0.81	0.15	0.70	0.91	0.64	0.74	0.85	0.09
4	0.11	0.85	0.56	0.94	0.79	0.66	0.19	0.11	0.18	0.58
5	0.71	0.99	0.65	0.29	0.74	0.93	0.67	0.89	0.60	0.43
6	0.70	0.63	0.64	0.25	0.13	0.53	0.25	0.42	0.45	0.44
7	0.47	0.62	0.84	0.98	0.59	0.59	0.03	0.37	0.29	0.58
8	0.99	0.58	0.27	0.74	0.95	0.50	0.94	0.26	0.43	0.43
9	0.11	0.78	0.40	0.99	0.57	0.50	0.05	0.43	0.28	0.95
10	0.52	0.03	0.88	0.53	0.51	0.53	0.32	0.53	0.92	0.89
11	0.45	0.83	0.66	0.39	0.16	0.84	0.67	0.34	0.83	0.31
12	0.06	0.54	0.91	0.19	0.45	0.45	0.93	0.74	0.43	0.75
Total	5.23	7.63	7.07	7.01	6.70	6.89	6.35	6.10	6.54	5.83
Subtract 6	−0.77	1.62	1.07	1.01	0.70	0.89	0.35	0.10	0.54	−0.17
Times 0.5	−0.38	0.81	0.53	0.51	0.35	0.45	0.18	0.05	0.27	−0.08
Add 2	1.62	2.81	2.53	2.51	2.34	2.45	2.18	2.05	2.27	1.92
Exponential	5.03	16.7	12.6	12.3	10.5	11.5	8.82	7.77	9.69	6.80

4. Add μ, which changes the mean to μ and leaves the standard deviation unchanged.

This is illustrated in Table 2.14, for $\mu = 2$ and $\sigma = 0.5$, which shows the simulated normal values in the next to last line.

Log-normal As a variable is log-normal if its natural log is normal, generate a normal variable, and take its exponential (i.e. its anti-log). Table 2.14 also shows the simulated log-normal values in the last line.

Exercise 2.6

2.6.1 For each of the distributions binomial, Poisson, exponential, Erlang, and normal, generate a sample of 10 numbers. Calculate the mean and variance in each case, and see that they are approximately equal to the theoretical values. If a class or tutorial group is doing this exercise, combine the results, and if enough results can be generated by a class, draw the histograms and compare them with the pdfs.

Reliability activities at the design stage

3

3.1 Introduction

It is essential that reliability is considered at all stages of a project. This cannot be overemphasized or mentioned too often. In particular, when the designer starts his work on the drawing board, or even earlier on the back of a cigarette packet, he must consider reliability. The techniques presented here are design aids, in that if the analyses are done properly, at the appropriate time, and by the correct people, and the documentation is kept up to date, then the possible design weaknesses have a far higher probability of being detected and dealt with early in the project. Also management and/or the customer can receive assurance that the reliability is being properly addressed, and that there is a greater chance that the reliability will be high and/or the reliability specification will be met. It is also the case that by dealing with potential reliability problems at the design stage, the development time will be shorter, the corresponding costs will be less, and the reliability will be higher. A design change that costs one pound to make on the drawing board will cost a hundred pounds to incorporate into a prototype, and could cost several thousand pounds if it is necessary when the equipment is commissioned and in service.

Only three techniques are presented in this book, namely Failure Modes Effects (and Criticality) Analysis, Fault Tree Analysis, and Event Tree Analysis, or Cause Consequence Analysis as it is sometimes known. These techniques are very commonly used as design aids and for assurance purposes. They can also be inputs into design reviews, along with other reports, predictions, data analyses, etc.

3.2 Failure Modes Effects and Criticality Analysis

3.2.1 Introduction

Failure Modes Effects Analysis (FMEA) and Failure Modes Effects and Criticality Analysis (FMECA) are engineering analyses that, if performed properly, and at the correct time, can be of great value in assisting the decision making process of the engineer during design and development. The analysis is often called a 'bottom up' analysis, as it examines low level components or functional groupings of components, and considers the systems failures that result from all their different failure modes. The term component will be used throughout this section to mean the lowest level grouping that is being considered in the analysis.

There are a number of standards dealing with this technique. The British standard BS 5760 gives an outline in part 2, *Guide to the assessment of reliability*, and Part 3, *Guide to reliability practices: examples*, gives an example. Part 5, *Guide to failure modes effects and criticality analysis (FMEA) and (FMECA)* gives guidance in the application of the techniques. The American military standard, US MIL STD 1629, (*Procedures for performing a failure modes effects and criticality analysis*) is considered by many to be the standard reference.

The reasons for doing a FME(C)A are many, but among the benefits are that:

- It provides designers with an understanding of the structure of the system, and the factors which influence reliability.
- It helps to identify items that are reliability sensitive or of high risk, and so gives a means of deciding priorities for corrective action.
- It identifies where special effort is needed during manufacture, assembly or maintenance.
- It establishes if there are any operational constraints resulting from the design.
- It gives assurance to management and/or customers that reliability is being or has been properly addressed early in the project.

A good FME(C)A will present, in tabular form, for each failure mode of all the components:

- the effect on sub-assembly, assembly, sub-system etc to system level;
- likelihood of occurrence;
- severity and criticality (for a criticality analysis).

The report may also include further information, such as recommendations for:

manufacture and assembly (inspection, test, quality of components etc.);

maintenance (inspection, test, replacement periods etc.);

detectability (user, maintainer, BITE etc.).

The information in an FMECA can also be used in the design of built in test, and is an essential input into Logistic Support Analysis and Reliability Centered Maintenance.

An FME(C)A may be quantitative, or qualitative, or a mixture of the two. Both approaches are described here. Likelihood of occurrence is

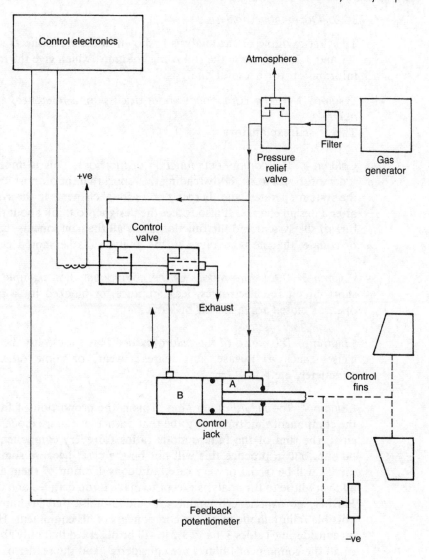

Fig. 3.1 Schematic diagram of hydraulic control system.

normally (though not always) quantified by the failure rate, and by assigning a numerical value to the severity of the effect of each failure mode on the system, the criticality of the failure modes can be measured by combining these values. High criticalities indicate areas for redesign or attention, either of the failure mode in question, or of a component or subassembly. The technique is best illustrated by way of examples, and Fig. 3.1 shows a hydraulic control system, that will be used to illustrate the techniques and methods discussed in the rest of the chapter.

3.2.2 Quantitative analysis

The first example of the analysis is shown in the work sheets in Tables 3.1 and 3.2. They have the following columns, which give the minimum information for a useful analysis.

Columns 1–3 The component under discussion, a reference, and code number
This is self explanatory.

Column 4 The components function or functions This is important as it demonstrates to anybody reading the report that the authors know how the system operates, and there are no residual elements in the system left after a design change. It also forces the designer to think about the structure of the system and the functioning of all the components, assuming, of course, that he is involved in the analysis (as he should be).

Column 5 The failure modes of the component For example open or short circuit for electronics, leak, rupture, or blocked for a pipe, fail open or closed for a switch or valve, etc.

Column 6 The cause of the failure mode This should list the possible causes, such as misuse, dirt ingress, wear, or some other ageing parameter, etc.

Column 7 The failure mode ratio This is the proportion of failures of the component that turn out to be that particular failure mode. In principal, the sum of the failure mode ratios for every component should be one, but in practice this will not be the case, because some failure modes will be trivial or very rare, and consideration of them all would add very little to the analysis except to make it too detailed and complex. Further, no experienced engineer would promise that he knows every possible failure mode of all the components in his equipment. However, on considering Tables 3.1 and 3.2 it will be observed that only about 65% of all the component failures are considered, and the reader of a report containing such worksheets could be forgiven for worrying that maybe some important failure modes had been omitted. In any case, a significant proportion of failures have clearly not been considered, and the assurance that could be gained from the report is not there. The failure mode ratio is denoted α.

Column 8 The failure rate of the component (not of the failure mode), denoted λ as usual The failure rate of the failure mode is given by the product, $\lambda\alpha$.

The source of the data for Columns 7 and 8 should also be stated somewhere in the report (not necessarily here), and justification for using

Table 3.1 Failure modes and effects analysis – worksheet

System: Missile
Subsystem: Control
Assembly: Actuator (C3)
Sub-assembly: Control Jack (C32)
Parts list no: A/1234/79

Item	Ref no	Code no	Function	Failure mode	Cause	Failure mode frequency (α)	Component failure rate (10^4 hrs) (λ)	Failure effect Immediate level*	Failure effect Next level*	Symptoms	Severity level (S)	Criticality (C) ($= \alpha.\lambda.\varsigma$)	Remarks
1	2	3	4	5	6	7	8	9	10	11	12	13	14
Cylinder block	CJ1	C321	Drive control fins	Restricted gas passages	Dirt ingress during manufacture	0.45	0.8	Reduced rate of piston travel	Worst case (blockage) no actuator drive		1.0	0.360	Safety hazard – loss of control
Cylinder block	CJ1	C321		Fracture	Misuse	0.30	0.8	Gas loss	Slow drive and drive will cease early	As in Column 9	0.7	0.168	
Cylinder seal	CJ2	C322	Prevent gas leakage	Leakage	Age	0.90	1.0	Gas loss	As above – slow drive		0.7	0.630	
Piston	CJ5	C323	Drive control fins.	Sticking	Lack of lubricant, dirt	0.60	1.5	Intermittent piston movement	Intermittent drive		0.7	0.630	
Piston	CJ5	C323		Seized	Lack of lubricant, misuse	0.15	1.5	No piston movement	No drive		1.0	0.225	Safety hazard loss of control
Piston seal	CJ6	C324	Prevent leakage across piston	Leakage	Age	0.8	0.7	Gas leakage across piston	Slow drive		0.4	0.224	
Connecting rod	CJ4	C325	Drive control fins	Bending	Misuse	0.65	0.5	Restricted or no movement	Slow or no drive		1.0	0.325	Possible safety hazard – loss of control

Total criticality 2.562

* Immediate level for this analysis   Subassembly
* Next level for this analysis   Assembly

Table 3.2 Failure modes and effects analysis – worksheet

System   Missile
Subsystem   Control
Assembly   Actuator (C3)
Sub-assembly   Control Valve (C32)
Parts list no   A/1234/79

Item	Ref no	Code no	Function	Failure mode	Cause	Failure mode frequency (α)	Component failure rate (10^6 hrs) (λ)	Failure effect Immediate level*	Failure effect Next level*	Symptoms	Severity level (S)	Criticality (C) ($= \alpha.\lambda.c$)	Remarks
1	2	3	4	5	6	7	8	9	10	11	12	13	14
Solenoid	CV6	C311	Operates valves	Open winding	Misuse, age	0.60	1.50	Inlet valve permanently open under gas pressure	Actuator 'hardover' right	As in Column 9	1.0	0.90	Safety hazard loss of control
Solenoid	CV6	C311		Insulation	Age	0.30	1.50				1.0	0.45	
Compression spring	CV10	C312	Open exhaust vv-close inlet vv	Fracture	Misuse	0.70	0.1	Reduced force to operate vv	Slower actuator drive		0.4	0.028	
Inlet and exhaust valves	CV5	C313	Meter Gas	Sticking	lack of lubricant	0.40	3.0	Incorrect metering	Slower actuator drive		0.4	0.48	
	CV5	C313	Flow	Degraded vv seats	Age	0.50	3.0				0.4	0.60	
Valve body	CJ7	C314	Meter Gas Flow	Restricted gas passages	Dirt ingress during manufacture	0.45	0.7	Incorrect metering	In worse case (ie blockage no drive or 'hardover' left. Slow actuator drive and drive will cease early		1.0	0.315	Safety hazard – loss of controls
Valve body	CV7	C314		Fracture	Misuse	0.40	0.7	Gas loss			0.7	0.196	

Total criticality   2.969

* Immediate level for this analysis   Sub-assembly
* Next level for this analysis   Assembly

it. It is a commonly quoted maxim that reliability in one system should not be inferred from data from another source; but there are times when no data exists, as when trying to assure reliability before any prototypes are built, as could well be the case here. In such a situation, the database used should be justified, and possible differences in environment, that could lead to the data being poor for the application concerned, highlighted. If engineering judgment is used, either to obtain a failure rate or failure mode ratio figure, or to adjust either values that come from a database, then that, too, must be highlighted.

Columns 9–10 The effects of the failure on the immediate level (subassembly) and the next level (assembly) For a complex system, such as an aircraft, the effect on several different levels should be considered. In particular, the effect on the total system, and the mission(s) or task(s) should be shown. There may also be a necessity to show the effect on different phases of the equipment use.

So, for example, if the washer on a tap wears, the tap may drip, which can be annoying, but it will not impair the flow of water when the tap is turned on.

Column 11 Symptoms These should be different from, or add to, anything in Columns 8 and 9, and should aid the user in diagnosing faults before they become serious, or have a serious effect, if at all possible. If there are no symptoms until a system failure occurs, this should be highlighted, and a possible design change considered.

Column 12 The severity level A simple definition of severity is the proportion of times that the failure mode in question will cause system failure. In the absence of sufficient data, this may be the result of engineering judgment, but a greater difficulty than the absence of data is that it does not take into account the system failure mode, the risk to the system, the possible degree of damage and financial loss, and the risk of injury or maybe even death to the operator and other personnel. Severity should really be measured using two parameters, one of which is a measure of the functionality of the system, and the other taking into account any other effects. A subjective value, that combines both factors, is usually acceptable, as long as the rules for deciding the values of the severity level are decided in advance, and standardized, and agreed by all concerned. An example of the definition of severity levels is given in Table 3.3. There may be advantages in the qualitative approach described below. The severity level is denoted S.

Column 13 The criticality The criticality of a failure mode is the combination of the effect of that failure mode on the system and its

Table 3.3 Example of severity scale of values

Scale value	Description
1.0	A failure resulting in system loss or high probability of death or serious injury.
0.7	A failure that results in loss of functionality or injury to the operator.
0.4	A failure that results in degraded performance.

frequency of occurrence. In this case, the criticality of each failure mode is given by

$$C_m = \lambda \alpha S$$

and the criticality of each component is given by

$$C_c = \sum C_m$$

summed over all the failure modes of each component. The criticality of the subassembly, which is sometimes, though infrequently, required, is given by

$$C_s = \sum C_c$$

summed over all the components of the subassembly.

Column 14 Remarks This column is reserved for any further comments that the authors may wish to make, and any points that they feel should be brought to the attention of management, the authorities, or a potential customer. In particular, any potential safety related features can be highlighted. (An FMECA is frequently done as part of a safety case, and is acceptable by most bodies concerned with safety, such as the HSE. Safety is not reliability, however, despite the fact that there is some overlap in the theory. In some instances, safety and reliability pull in opposite directions.) It can also show something about the structure of the overall system. The comments here all relate to a loss of control, which is not surprising as the system being analysed is a control system. If an analysis of the air-conditioning in an airliner showed that there was a risk of the doors opening, to take an extreme example, one could ask some very telling questions of the design authority.

Consider for example, the control valve. The valve body has a failure rate of 0.7 failures/10^6 hr. Of these failures, 45 percent are restricted gas passages and 40 percent are fracture. The remaining 15 percent are not analysed. The effect of these is that for the former the metering will be hard to correct, leading in the worst case to no drive with the fins 'hard over', and loss of control, while the latter will cause a slow actuator drive, and the drive will probably cease early. The severity levels are 1.0 and 0.7 respectively, leading to criticalities of

$$0.45 \times 0.7 \times 1.0 = 0.315$$

and

$$0.4 \times 0.7 \times 0.7 = 0.196,$$

with a total criticality of the valve body of 0.511.

This figure of 0.511 is low compared with the criticalities of other components, e.g. the solenoid criticality is 1.35, and so it has a low priority when considering design changes. Should it become subject to further development, then the gas passages problem is more important than fracture.

The total criticality of the control valve is 2.969, slightly worse than that of the control jack at 2.562. This difference is too small to be significant, and the presentation of this result could be criticized for being given to three places of decimals, particularly when the source of the data in the worksheets is considered. Although FME(C)A is a powerful tool, the numerical, predictive part is no more accurate than any other technique, three places of decimals cannot be justified, and could give the quantitative analysis a respectability it does not deserve.

On the report, the following points should also be noted:

- a full definition of the system, its function and of failure;
- the levels of assembly, subassembly etc.;
- all the components or functional groupings should be clearly identified;
- the environment and operational stresses should be clearly detailed;
- the source and reliability of all the data should be stated.

3.2.3 Severity classification

An alternative approach to the measure of severity given above is to put the severities into one of a number of categories. Table 3.4 shows a system of four categories which is given as an example.

These definitions are not intended to be definitive, but are presented for illustrative purposes only. There may also be more than four categories if it is thought necessary, and they may be numbered rather than lettered. However, if they are numbered, there is a danger that the value may be considered a measure of severity (which may not be a bad thing). It is important that if a numerical system is used, that the scales of values used all measure in the same direction. Failure rate, for example, is a measure of badness rather than goodness, and so to avoid

Table 3.4 Example of severity classification

Category and type	Description
A *Minor*	A failure which would require unscheduled maintenance.
B *Major*	A failure that may cause minor injury or damage, or which may delay the mission.
C *Critical*	A failure that could cause severe injury or damage, and that would cause mission failure.
D *Catastrophic*	A failure that may cause death or system loss.

confusion, high numbers used in the measure of severity or criticality should indicate a worse situation than low numbers.

Tables 3.5 and 3.6 show an FMECA done using this qualitative approach to severity. The categories used are those quoted above. The columns used are similar to those used in the first example, with the following exceptions:

7a Failure rate This is the failure rate of the failure mode under consideration, and is $\lambda\alpha$ in the notation of the first example. This is an alternative way of presenting this data. It has the advantage of making the analysis simpler, but the disadvantage of omitting consideration of the failure mode ratio, and hence of consideration of the fraction of failures that are not considered in the analysis.

12a The severity level This is as defined above, and takes one of the values A–D.

There is no Column 13.

3.2.4 Summary measures of criticality

There is no way of producing a summary measure of the criticality of each component or subassembly as there was with the previous technique. Two pictorial methods are presented here.

Figures 3.2 and 3.3 show criticality histograms for the valve and the jack. The horizontal axes are the severity classification, and the vertical axes are failure rate. The height of each bar in the histogram is equal to the sum of the failure rates of each failure mode with severity indicated by the position on the horizontal axis. Clearly, the higher the bars to the right of the chart, the more cause for alarm.

Figures 3.4 and 3.5 show criticality matrices for the two subassemblies. The axes are the same as the ones described above, but now each failure mode is represented by a point on the chart, showing its criticality and failure rate. The horizontal axis could be the scale from

Table 3.5 Failure modes and effects analysis – worksheet

System   Missile
Subsystem   Control
Assembly   Actuator (C3)
Sub-assembly   Control Jack (C32)
Parts list no   A/1234/79

Item	Ref no	Code no	Function	Failure mode	Cause	Failure mode Failure rate	Failure effect Immediate level*	Failure effect Next level*	Symptoms	Severity level (S)	Remarks
1	2	3	4	5	6	7a	9	10	11	12a	14
Cylinder block	CJ1	C321	Drive control fins	Restricted gas passages	Dirt ingress during manufacture	0.36	Reduced rate of piston travel	In worst case (ie blockage) no actuator drive		D	Safety hazard – loss of control
Cylinder block	CJ1	C321		Fracture	Misuse	0.24	Gas loss	Slow drive and drive will cease early		C	
Cylinder seal	CJ2	C322	Prevent gas leakage	Leakage	Age	0.90	Gas loss	As above – slow drive		C	
Piston	CJ5	C323	Drive control fins	Sticking	Lack of lubricant, dirt	0.90	Intermittent piston movement	Intermittent drive	As in Column 9	B	
Piston	CJ5	C323		Seized	Lack of lubricant, misuse	0.23	No piston movement	No drive		D	Safety hazard loss of control
Piston seal	CJ6	C324	Prevent leakage across piston	Leakage	Age	0.56	Gas leakage across piston	Slow drive		B	
Connecting rod	CJ4	C325	Drive control fins	Bending	Misuse	0.33	Restricted or no movement	Slow or no drive		D	Possible safety hazard – loss of control

* Immediate level for this analysis   Subassembly
* Next level for this analysis   Assembly

Table 3.6 Failure modes and effects analysis – worksheet

System — Missile
Subsystem — Control
Assembly — Actuator (C3)
Sub-assembly — Control Valve (C32)
Parts list no — A/1234/79

Item	Ref no	Code no	Function	Failure mode	Cause	Failure rate	Failure effect Immediate level*	Failure effect Next level*	Symptoms	Severity level (S)	Remarks
1	2	3	4	5	6	7a	9	10	11	12a	14
Solenoid	CV6	C311	Operates valves	Open winding	Misuse, age	0.90	Inlet valve permanently open under gas pressure	Actuator 'hard-cover' right	As in Column 9	D	Safety hazard – loss of control
Solenoid	CV6	C311		Insulation	Age	0.45				D	
Compression spring	CV10	C312	Open exhaust vv - close inlet vv	Fracture	Misuse	0.07	Reduced force to operate vv	Slower actuator drive		B	
Inlet and exhaust valves	CV5	C313	Meter gas	Sticking	Lack of lubricant	1.20	Incorrect metering	Slower actuator drive		B	
	CV5	C313	flow	Degraded vv seats	Age	1.50				B	
Valve body	CV7	C314	Meter gas	Restricted gas passages	Dirt ingress during manufacture	0.32	Incorrect metering	In worst case (ie blockage) no drive or 'hardcover' left		D	Safety hazard – loss of controls
Valve body	CV7	C314	flow	Fracture	Misuse	0.28	Gas loss	Slow actuator drive and drive will cease early		B	

* Immediate level for this analysis Subassembly
* Next level for this analysis Assembly

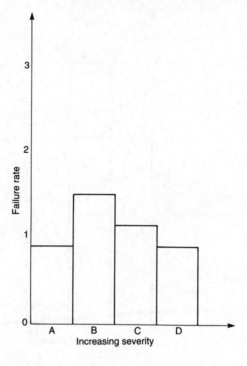

Fig. 3.2 Control jack criticality histogram.

zero to one used in the first example, rather than the A–D scale used in the second, but the picture would be very similar. Points high up on the left of the matrix are frequent, trivial things going wrong, which would annoy rather than impair operation, while points low down on the right are serious, infrequent failures. If there is anything high and to the right, the equipment should never be built, as it is totally unsatisfactory. Redesign can move points down the matrix relatively easily, but it is much harder to move them to the left: i.e. it is more difficult to make failures less severe in their effect than it is to make them less frequent.

3.2.5 Qualitative

A third technique, completely qualitative, and commonly used in industry, is to extend the idea of categorization to all other measures that are being considered, as well as severity. Tables 3.7, 3.8, and 3.9 define 10 categories of failure probability, detectability and severity based on an industrial standard used when doing FMEAs on a manufacturing process. In practice, these three numbers were multiplied together to produce the risk priority number (RPN) for each failure mode. The RPN

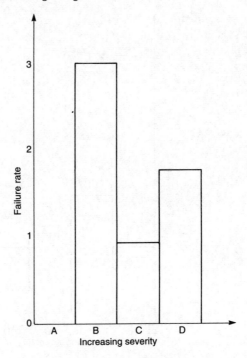

Fig. 3.3 Control valve criticality histogram.

plays much the same role as criticality. The high values of the RPN are 'designed out', usually by improving the quality of the component concerned or by improving the detectability, as the severity is generally difficult to reduce (though not impossible, if redundancy, for example, is considered a viable option).

The advantage of this method is that there is no prolonged debate over trivial details, such as the nth place of decimals, but is rather a comparative technique, in which the analyst can compare failure modes with previous results, the reasoning being along the lines of 'this is worse than X, which rated 5, but not as bad as Y which rated 8, so 6 or 7 seems right, and Z was rated 6, and its about the same as Z, so give it 6'. This subjective approach can very often be used early in the project, before any trials data are available, and the FME(C)A changed to a quantitative one as data are generated.

A team performing an FME(C)A and using this approach is free to alter the definitions in these tables as it sees fit, although it must be consistent throughout the analysis. The same comment applies to the number of categories, but more than 10 is far too many, and fewer than 4 may give too coarse a resolution. The scale of ratings should be defined

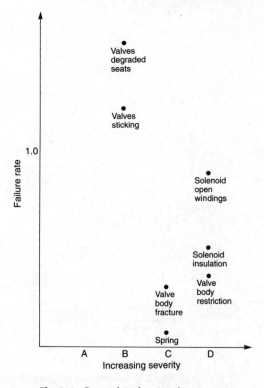

Fig. 3.4 Control jack criticality matrix.

in such a way that the greatest resolution is obtained in the region of greatest interest, i.e. where most of the component data lies.

Tables 3.10 and 3.11 are repeats of the previous FMECAs on the hydraulic control system using this technique. Note that the RPN includes more information than the measures considered earlier, and so the results are not directly comparable.

3.2.6 Further considerations

FME(C)A can be a very powerful tool, for identifying design weaknesses, indicating areas for further development and giving assurance to the project manager etc., if it is done properly. Ideally it is a team effort, involving at the very least the designers, with help from other interested parties, such as maintenance and sales, including expert assistance from a specialist reliability group. If the designer's logs are properly kept, they should form the basis of the first report. It should also be amended regularly as the design is changed, and so is a living document, being up to date with all the design changes and reflecting

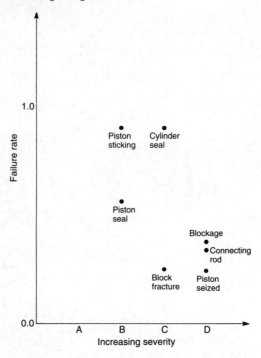

Fig. 3.5 Control valve criticality matrix.

the current status of the system. Like all supporting documentation, it is of far more value if it is updated when any changes are made, rather than being written once at the end of the project (to take the extreme case). The practice of some contractors of handing the design over to a contractor at the end of the project in order to fulfill a contractual obligation, which is so often done that the FME(C)A is nothing more than a deliverable and is no more than an expensive, useless exercise that wastes the customer's money.

There are other possible outputs from an FMECA, including recommendations for preventative maintenance, design, and auditing of built in test equipment (BITE). An FMECA is a necessary input into Reliability Centered Maintenance (RCM) and Logistic Support Analysis (LSA).

Exercise 3.1

3.1.1   It is difficult to write traditional exercises on FME(C)As, and it is suggested that the students form into small groups, of five or six, in order to carry out an FME(C)A on a simple system. Suggestions for the subject of the analysis are:

a water tap;

an electric plug or socket

Table 3.7 Example of categories of likelihood of occurrence

Criteria	Rating	Possible failure rate
Remote probability of occurrence. It would be very unlikely for these failures to be observed even once.	1	0
Low probability. Likely to occur once, but unlikely to occur more frequently.	2 3	1:20 000 1:10 000
Moderate probability. Likely to occur more than once.	4 5 6	1:2000 1:1000 1:200
High probability. Near certain to occur at least once.	7 8	1:100 1:20
Very high probability. Near certain to occur several times.	9 10	1:10 1:2

Table 3.8 Example of categorization of severity

Criteria	Rating
Minor. A failure that has no effect on the system performance. The operator would probably not notice.	1
Low. A failure that would cause slight annoyance to the operator, but that would cause no deterioration to the system.	2 3
Moderate. A failure that would cause a high degree of operator dissatisfaction (i.e. high pedal effort, or radio buzz), or that causes noticeable, but slight deterioration in system performance.	4 5 6
High. A failure that causes significant deterioration in system performance, but that does not effect safety.	7 8
Very High. A failure that which would seriously effect the ability to complete the task or which could cause damage, serious injury or death.	9 10

a hydraulic valve;

a car distributor;

a simple piece of electronics, or a single chip (maybe an 'OR' or an 'AND' gate), or a transistor.

In the last example of Exercise 3.1.1, it would be possible for the output from a FME(C)A on a transistor to be the input to one on a gate. The list is by no means exhaustive, and need only depend on the skills and experience of the students and the imagination of the lecturer.

Table 3.9 Example of categorization of detectability

Criteria	Rating	Probability (%)
Remote probability that the failure remains undetected. Such a defect would almost certainly be detected during inspection or test.	1	86–100
Low probability that the defect remains undetected	2	76–85
	3	66–75
Moderate probability that the defect remains undetected	4	56–65
	5	46–55
	6	36–45
High probability that the defect remains undetected	7	26–35
	8	16–25
Very high probability that the defect remains undetected until the system performance degrades to the extent that the task will not be completed.	9	6–15
	10	0–5

3.3 Fault Tree Analysis

3.3.1 Introduction

Fault Tree Analysis (FTA) is a technique used in the analysis of complex systems that has been around for many years. It complements FMECA in that it is a 'top-down' analysis, starting with a system fault (the top event), and analysing this fault in terms of subsystem faults. This is illustrated in Fig. 3.6, which shows the preliminary analysis of a calculator, where the top event is the fault 'no liquid crystal display'. This can be caused either because the LCD has failed, or because there is no power reaching the display. The latter fault is analysed further, and may be caused by the faults shown, either 'switch failed at off', or 'internal wiring failure', or 'no power to the calculator'. These failures can be analysed yet further as shown in Fig. 3.6.

The benefits of performing an FTA can be summarized as follows. First, at the simplest level it imposes a discipline on the designer (assuming he is involved in the analysis, which he should be), by forcing him to think of the system in terms of how it may fail, and how the subsystems interact. It also gives insight into the structure of the system, and any activity that does these things can only be of benefit, and give assurance to the customer or user of the system. The system structure can also be analysed further in terms of the minimum cut sets of the system. The analysis has a powerful mathematical foundation in Boolean algebra, which will not be studied here. The interested reader is referred

Table 3.10 Failure modes and effects analysis – worksheet

System	Missile
sub-system	Control
assembly	Actuator (C3)
Subassembly	Control Jack (C32)
Parts list no	A/1234/79

						Scale values			Failure effect			Risk priority number	Remarks
Item	Ref no	Code no	Function	Failure mode	Cause	Rate of occurrence	Severity	Detectability	Immediate level	Next level	Symptoms		
1	2	3	4	5	6	7(c)i	7(c)ii	7(c)iii	9	10	11	12	14
Cylinder block	CJ1	C321	Drive control	Restricted gas passages	Dirt ingress during manufacture	6	10	6	Reduced rate of piston travel	In worst case (ie blockage) no actuator drive.	As in Column 9	360	Safety hazard-loss of control
Cylinder block	CJ1	C321	fins	Fracture	Misuse	6	6	3	Gas loss	Slow drive and drive will cease early		108	
Cylinder seal	CJ2	C322	Prevent gas leakage	Leakage	Age	7	6	8	Gas loss	As above-slow drive		336	
Piston	CJ5	C323	Drive control	Sticking	Lack of lubricant, dirt	7	4	3	Intermittent piston movement	Intermittent drive		84	
Piston	CJ5	C323	fins	Seized	Lack of lubricant, misuse	5	10	3	No piston movement	No drive		150	Safety hazard loss of control
Piston seal	CJ6	C324	Prevent leakage across	Leakage	Age	6	5	8	Gas leakage across piston	Slow drive		240	
Connecting rod	CJ4	C325	Drive control fins	Bending	Misuse	5	10	2	Restricted or no movement	Slow or no drive		100	Possible safety hazard loss of control

* Immediate level for this analysis   Subassembly
* Next level for this analysis   Assembly

Table 3.11 Failure modes and effects analysis – worksheet

System   Missile
Subsystem   Control
Assembly   Actuator (C3)
Subassembly   Control valve (C31)
Parts list no   A/1234/79

						Scale values 7c			Failure effect				
Item	Ref no	Code no	Function	Failure mode	Cause	Rate of occurrence	Severity	Detectability	Immediate level*	Next level*	Symptoms	Risk priority number	Remarks
1	2	3	4	5	6	7(c)i	7(c)ii	7(c)iii	9	11	11	12	14
Solenoid	CV6	C311	Operates valves	Open winding	Misuse, age	7	10	2	Inlet valve permanently open under gas pressure	Actuator 'hardcover' right		140	Safety hazard-loss of control
Solenoid	CV6	C311		Insulation	Age	6	9	3				162	
Compression spring	CV10	C312	Open exhaust vv – close inlet vv	Fracture	Misuse	4	4	2	Reduced force to operate vv	Slower actuator drive	As in Column 9	32	
Inlet and exhaust valves	CV5	C313	Meter	Sticking	Lack of lubricant	7	3	3	Incorrect metering	Slower actuator		63	
Inlet and exhaust valves	CV5	C313	gas flow	Degraded vv seats	Age	7	3	3	Incorrect metering			63	
Valve body	CV7	C314	Meter gas	Restricted gas passages	Dirt ingress during manufacture	6	10	6	Incorrect metering	In worst case (ie blockage) no drive or 'hardover' left. Slow actuator		360	
Valve body	CV7	C314	flow	Fracture	Misuse	6	4	3	Gas loss	drive and drive will cease early		72	Safety hazard – loss of controls

* Immediate level for this analysis   Subassembly
* Next level for this analysis   Assembly

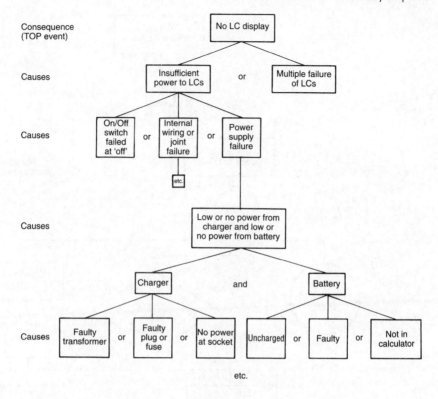

Fig. 3.6 Liquid crystal display informal fault tree.

to Pagés and Gondran (1986). The analysis can also include potentially fatal events that are not necessarily of a mechanical nature, such as lightning strikes, or as is illustrated in the example of the calculator, human activities. One of the basic failures is 'no battery in the calculator'. This cannot be due to wear or any other mechanical cause, but must be because the battery has been removed by the operator, or somebody else, and so is a human induced failure. Lastly, if numerical data is available, the probability of failure (the top event) can be calculated, as explained below.

3.3.2 Analysis

Fig. 3.6 shows a preliminary analysis only. The formal analysis is shown in Fig. 3.7, using the symbols that are explained in Table 3.12. Because of the logical nature of the system structure, a fault tree is a pictorial representation of a Boolean statement, and Boolean theorems can be used to analyse the tree. In particular, all analysis can be done using OR gates and AND gates.

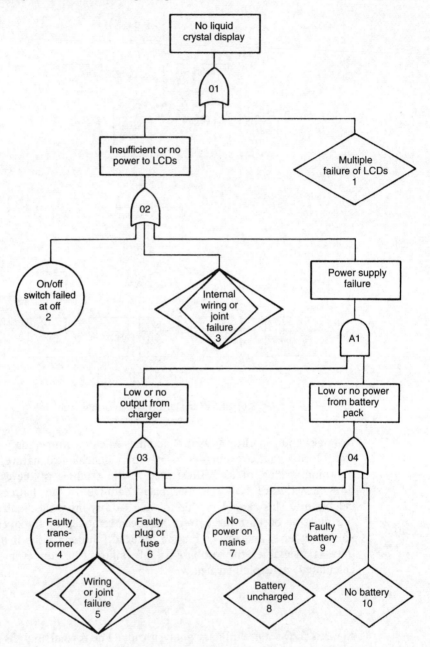

Fig. 3.7 Liquid crystal display fault tree.

Table 3.12 Symbols used in Fault Tree Analysis

Symbol	Description and use
	PRIMARY BASIC EVENT. A fault that is not analysed further. The failure probability can be derived from empirical data.
	SECONDARY BASIC EVENT. A fault that could be further analysed, but this is not done through lack of information or other reasons.
	SECONDARY BASIC EVENT. A fault that could be analysed further, but this will not be done until later.
	OUTPUT EVENT. An event that is analysed, and is the result of combinations of other events through a logic gate
	AND GATE. The output event (next higher event) will occur only if all the input events occur.
	OR GATE The output event (next higher event) will occur if any one of the input events occurs.

In simple cases (i.e. where only OR and AND gates are used and where no failure occurs more than once in the tree) if numerical failure data is available, then the following rules can be used to calculate the probability of the top event happening.

The probability of the output to an OR gate is approximately equal to the sum of probabilities of the input failures, and the probability of the output to an AND gate is equal to the product of probabilities of the input failures.

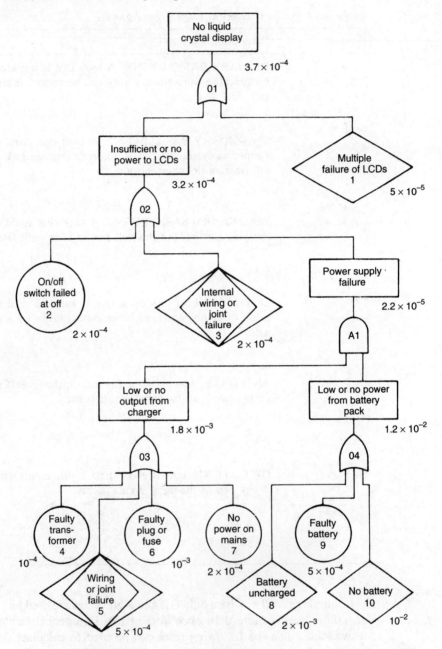

Fig. 3.8 Liquid crystal display fault tree: quantitative analysis.

The use of these rules is illustrated in Fig. 3.8 using the LCD example.

3.3.3 Min. cuts

A cut set is a set (or collection) of component failures that will produce system failure, and a *minimum cut set*, abbreviated to min. cut, is a cut set that has no subsets that are cut sets. In plain language this means that a min cut is the is the limiting case of a cut set, i.e. it is a set of component failures that will produce system failure, with the further property that any single repair is sufficient to bring the system back to the functioning condition again. A list of all the min. cuts of a system tells the designer or user all there is to know about the reliability structure of the system, and if the data are available, the probability of failure of the system can be calculated. For example, the cut sets of the LCD example are given in Table 3.13, along with the numerical analysis. The probability of occurrence of each cut set is the product of the failure probabilities of each component in the set, and the failure probability of the system is the sum of the probabilities of occurrence of each cut set.

It is possible to calculate the min. cuts from the tree using Boolean algebra. In practice, the algebra can be summarized in a simple algorithm. The algorithm is easily implemented on a computer, and commercial fault tree analysis software should be able to calculate the min. cuts. The algorithm will not be described here, but the interested reader will be able to find it in Barlow and Proschan (1981).

Table 3.13 Min. cuts of LCD

Min cut	Probability
1	5×10^{-5}
2	2×10^{-4}
3	1×10^{-4}
4,8	2×10^{-7}
4,9	5×10^{-8}
4,10	1×10^{-6}
5,8	1×10^{-6}
5,9	2.5×10^{-7}
5,10	5×10^{-6}
6,8	2×10^{-6}
6,9	5×10^{-7}
6,10	1×10^{-5}
7,8	4×10^{-7}
7,9	1×10^{-7}
7,10	2×10^{-6}
System failure probability	3.7×10^{-4}

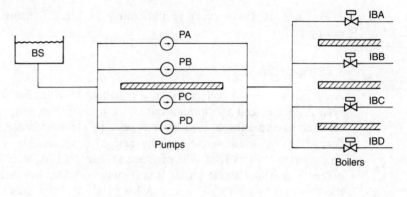

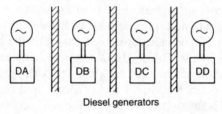

Fig. 3.9 Heat exchanger system.

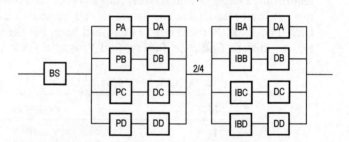

Fig. 3.10 Residual block diagram for heat exchanger system.

When a numerical result is required for more complex examples, the simple rules stated in Section 3.3.3 are not sufficient. Consider the system shown in diagrammatic form in Fig. 3.9. This is a heat exchanger system in a nuclear power generation plant. The heart of the system is the four heat exchangers, IBA to IBD. The system has some redundancy, and is considered to be operating if only two of the heat exchangers are functioning, but one heat exchanger alone is not sufficient (so the system is a 2-out-of-4 system). Each heat exchanger consists of a system of pipes and valves to control the flow of steam, and the valves are automatically controlled and electrically operated, obtaining their power from four diesel generators, DA to DD, each generator supplying power to just one heat exchanger. There are then essentially two failure modes for the heat

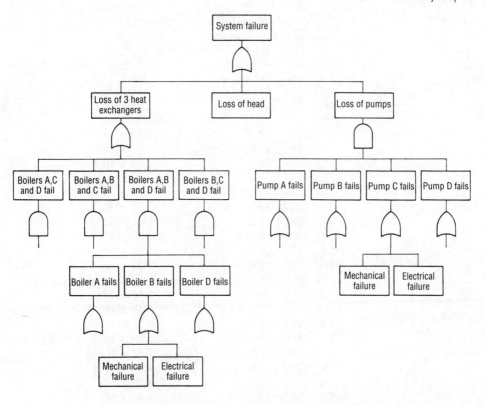

Fig. 3.11 Fault tree for heat exchanger system.

exchangers, either a failure in the exchanger itself, or a failure of the generator. These will be termed mechanical and electrical failures respectively. The water for the exchangers is pumped into them by the four pumps PA to PD. These are so configured and are of sufficient power that if any one pump is working it can serve any number of the exchangers. So the pump system is a 1-out-of-4 redundant system. The pumps are electric, and also take their power from the diesel generators DA to DD, each generator serving just one pump. As with the heat exchangers, the pumps can suffer either a mechanical or an electrical failure. Note that the generators each serve two purposes, supplying power to a heat exchanger and a pump. The RBD for this system is shown in Fig. 3.10.

The fault tree for this system is shown in Fig. 3.11. Note that, in order to make the point, the 2/4 redundancy of the heat exchangers is constructed using only OR and AND gates (any reasonable software would have '2/4 redundancy' gates), and that a number of components appear in more than one place in the tree. It is for that reason that the simple rules given for combining reliabilities used in the previous example will

Table 3.14 Min. cuts and probabilities for the heat exchanger system

DD	DC	DB		1.25×10^{-4}
DD	DC	DA		1.25×10^{-4}
DD	DB	DA		1.25×10^{-4}
DC	DB	DA		1.25×10^{-4}
BS				1.00×10^{-4}
DD	DC	IBA		7.50×10^{-5}
DD	DC	IBB		7.50×10^{-5}
DD	DB	IBC		7.50×10^{-5}
DD	DB	IBA		7.50×10^{-5}
DD	DA	IBB		7.50×10^{-5}
DD	DA	IBC		7.50×10^{-5}
DC	DB	IBD		7.50×10^{-5}
DC	DB	IBA		7.50×10^{-5}
DC	DA	IBD		7.50×10^{-5}
DC	DA	IBB		7.50×10^{-5}
DB	DA	IBD		7.50×10^{-5}
DB	DA	IBC		7.50×10^{-5}
DD	IBC	IBB		4.50×10^{-5}
DD	IBC	IBA		4.50×10^{-5}
DD	IBB	IBA		4.50×10^{-5}
DC	IBD	IBB		4.50×10^{-5}
DC	IBD	IBA		4.50×10^{-5}
DC	IBB	IBA		4.50×10^{-5}
DB	IBD	IBC		4.50×10^{-5}
DB	IBD	IBA		4.50×10^{-5}
DB	IBC	IBA		4.50×10^{-5}
DA	IBD	IBC		4.50×10^{-5}
DA	IBD	IBB		4.50×10^{-5}
DA	IBC	IBB		4.50×10^{-5}
IBA	IBB	IBC		2.70×10^{-5}
IBA	IBC	IBD		2.70×10^{-5}
IBA	IBB	IBD		2.70×10^{-5}
IBB	IBC	IBD		2.70×10^{-5}
DD	DC	PB	PA	1.0×10^{-6}
DB	DA	PD	PC	1.0×10^{-6}
DD	DB	PC	PA	1.0×10^{-6}
DC	DA	PD	PB	1.0×10^{-6}
DD	DA	PC	PB	1.0×10^{-6}
DC	DB	PD	PA	1.0×10^{-6}
DD	PC	PB	PA	4.00×10^{-7}
DC	PD	PB	PA	4.00×10^{-7}
DB	PD	PC	PA	4.00×10^{-7}
DA	PD	PC	PB	4.00×10^{-7}
PD	PC	PB	PA	1.60×10^{-7}
System failure probability				2.16×10^{-3}

3.3 Fault Tree Analysis | 93

Table 3.15 Failure data for the heat exchange system

Item	Symbol	Failure probability
Head Tank	BS	1×10^{-4}
Generator	D	5×10^{-2}
Pump	P	2×10^{-2}
Heat exchanger	IB	3×10^{-2}

not work in this case, as the events concerned are no longer independent, and a more refined tool is needed.

Table 3.14 lists the min. cuts of the heat exchanger system of Fig. 3.10. The failure data for this system are in Table 3.15, and Table 3.14 also summarizes the numerical analysis, and gives the numerical conclusion that the failure probability is 2.16×10^{-3}.

In more complex problems, the number of min. cuts can be very large, and in order to simplify the problem it is common to censor the number of cut sets by removing either those that are too long (i.e. have more than a critical number of failures in them), or those that are very unlikely, defined as those with a probability of occurrence that is less than some critical value. This leads to an optimistic estimate of the system failure probability. For example, in the heat exchanger system, if all the cut sets with a probability of occurrence of less than 10^{-4} are ignored on the grounds that they are too unlikely to be considered, then the estimate of the system failure probability is 6×10^{-4}, which is four times too small. In practice, there is no quantitative answer to this problem and engineering judgment must be used, but care must be taken to ensure that the baby is not thrown out with the bath water, and that the trivial many does not, in practice, overwhelm the important few.

3.3.4 Comparison of FTA and FMEA/FMECA

While FME(C)A is a bottom up approach, FTA is a top down analysis, and the two techniques complement each other. For a system with little or no redundancy, an FME(C)A is probably the best way, but for a complex system, with lots of back up, an FTA should probably be done. The last example, the heat exchanger system, illustrates this. There would be little point in continuing the FTA further in this case, as there is very little redundancy, certainly in the cases of the pumps and the generators, and a FME(C)A on these items would give the failure probabilities used in the FTA. (This is the point at which then FME(C)A coming up meets the FTA coming down.)

In doing either analyses, all those with knowledge or experience of the system should be involved, if it is at all possible. It is not uncommon to hand drawings over to a specialist contractor. While a fresh mind looking at the analysis is a good idea, getting a contractor performing all the analysis without any input from the designer or anybody else who may know the system (such as the user, the maintainer, or the producer) will not only be expensive, it will not give as much assurance nor be as useful.

As with FMECA there are standards dealing with FTA. BS 5760 Parts 2 and 3 have sections on FTA, while Part 7, *Guide to fault tree analysis*, is devoted to it.

Exercise 3.2 3.2.1 As with FME(C)A, it is difficult to write suitable exercises for FTA, as it is essentially a team activity. The class should be divided into small groups of about five or six, and set the task of performing an FTA on a relatively simple system. Suitable top events might be:

car spontaneously catches fire when ignition is off

tap fails to close

one of the outputs from an 'OR' or 'AND' gate is incorrect

Otherwise, similar comments apply here as were given when finding a suitable exercise for FME(C)A.

3.4 Event trees

Event trees are frequently used in the analysis of sequences of events, including human activities, that can lead to disasters or undesirable events. The activity is sometimes called cause consequence analysis, and is used more frequently in safety studies than reliability, but is included here for the sake of completeness.

Consider the following simple situation, in which it is considered necessary to study a fire alarm system. Ideally, if there is a fire, then the alarm goes off and a sprinkler system extinguishes the fire, but at each stage there is a human standby, in that if either the alarm or the sprinkler system fails a human operator can operate either or both. This can be represented by the event tree shown in Fig. 3.12.

Notice that of all the possible outcomes, to the right in Fig. 3.12, only three are that the fire spreads. The possible sequences of events that can lead to this undesirable event can now be identified from these outcomes, which are, after the fire starts:

The alarm fails to function and the operator fails to notice and take action in time.

The alarm functions but the sprinkler fails to function and the operator fails to notice and take action in time.

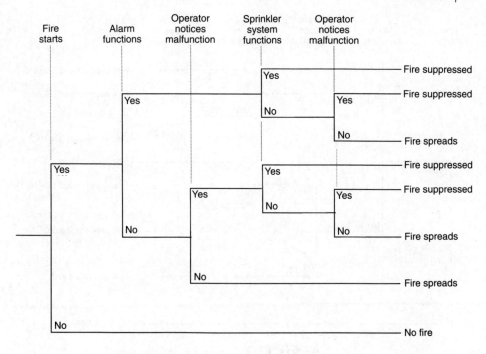

Fig. 3.12 Event tree.

The alarm fails to function, and the operator notices, but the sprinkler fails to function, and the operator fails to notice.

If sufficient data exist to estimate the probabilities, the likelihoods of the various outcomes can be obtained. This is shown in Fig. 3.13, from which it can be deduced that if a fire starts, then the probability of it spreading is given by:·

P(fire spreading) = P(alarm failing and operator not noticing)
+ P(alarm functioning and sprinkler not function-ing and operator not noticing and alarm failure and operator noticing and sprinkler failing and operator not noticing)
= 2.1 × 10⁻⁴.

Some of the intermediate steps are shown on the event tree in Fig. 3.13.

Notice that at each stage the branch that says 'yes' goes upwards, while the one that says 'no' goes down, in accordance with common conven-tion. Note too that time is passing as the tree goes from left to right. The probabilities may be obtained from trials, data, observation or judg-ment. Clearly the hardware failure probabilities can be obtained from data and/or modelling, but the human activity probabilities are more

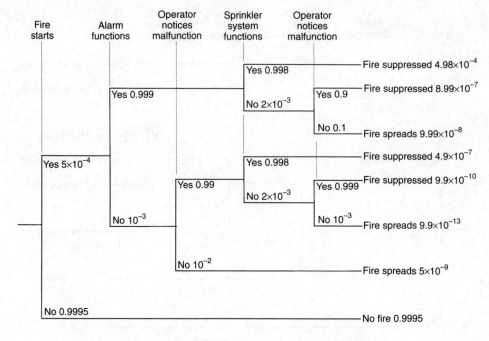

Fig. 3.13 Event tree: quantitative analysis.

difficult to obtain. This text does not discuss human factors, and the reader is directed towards other texts, such as Villemeur (1991), but sufficient to say here that for the purposes of assessing reliability and safety, probabilities are put on human activities.

This is only an outline of event tree analysis, and more can be found in Villemeur.

Exercises 3.3

Do an event tree analysis on each of the following:

3.3.1 Starting a car.

3.3.2 Lighting a gas fire with a cigarette lighter.

3.3.3 Cooking a pan of soup.

Reliability data

4

4.1 Introduction

Equipment is first designed on the drawing board, and then subsystems and eventually prototypes are built, and tested. During these tests failure data are generated, from the prototypes, their subsystems, and from components. This chapter describes some of the more widely known techniques used in the analysis of reliability and life data.

4.2 Reliability growth

4.2.1 Introduction

The question 'How does one make equipment reliable?' is often asked, to which the answer is often given: 'it must be designed in'. This is certainly true, and work that is done at the drawing board stage is always orders of magnitude cheaper than design changes that are incorporated into hardware. Despite this, all projects should have within them a reliability growth programme. This activity consists of testing prototype equipment, preferably in an environment at least as harsh as the one that the finished product will meet in practice, in order to make it fail, and hence find the design weaknesses. These areas of weakness can then be analysed and designed out: i.e. design changes can be incorporated to avoid the failure modes observed in the future. This technique is test, analyse, and fix, or test, find, and fix, or test, find, analyse and fix. Note that fix means a design fix, not a fix in the sense that the system is repaired to the same standard it was before the failure. It corresponds to debugging of software, and just as software always has bugs that need fixing, so hardware always has design weaknesses that need fixing.

In many situations it is necessary to quantify this process, in order to give assurance, or as an aid to making decisions concerning the length, cost, or effectiveness of a programme. There are a number of ways in which this can be done, of which the two commonest are described and discussed here, the Duane model and fix effectiveness factors.

Table 4.1 Data from a reliability growth programme

Failure number (n)	Cumulative time (hr)	M_c
1	103	103
2	315	157
3	801	267
4	1183	296
5	1345	269
6	2957	493
7	3909	558
8	5702	713
9	7261	807
10	8245	824

4.2.2 The Duane model

Consider the data of Table 4.1. It shows the cumulative time after the start of a reliability growth programme at which the failures were observed. So the first failure occurred after 103 hr, which may have been 103 hr on one prototype, or 51.5 hr on each of two prototypes, or if there were three prototypes one may have been running for 40 hr, another for 35 hr and the third for 28 hr when the failure happened. The second failure occurred after a total cumulative time of 315 hr, 212 hr after the first, and so on.

After the first failure the trials were assumed to have been suspended until the failure had been analysed, and a design change, that would hopefully reduce the probability of that particular failure mode happening again. The third column is the total test time divided by the number of failures, T/n, the MTBF, which is generally increasing, implying that the reliability is tending to increase. The MTBF is known as the cumulative MTBF and is denoted by M_c. The sort of question that may be asked after the 10th failure is the actual value of the MTBF. (Is it 824 hr, the last value in the table? One would think not, as this does not take account of the fact that the cause of the early failures appear to have been cured.) If the MTBF is not high enough, how many more resources are needed, and in particular, how much more time is needed to reach the target figure?

Duane (1965) noticed that in a reliability programme, if after development time T there had been n failures, then a plot of $\ln(T/n)$ against $\ln(T)$ was approximately linear, where ln is the natural logarithm. That is:

$$\ln(T/n) = \alpha\ln(T) + m \tag{4.1}$$

where α and m are constants.

The parameter α is called the growth factor, and is a measure of

Table 4.2 Comments on the effectiveness of a growth programme

$\alpha = 0.4\text{--}0.6$	Programme dedicated to the removal of design weakness and to reliability.
$\alpha = 0.3\text{--}0.4$	Well managed programme with reliability as a high priority.
$\alpha = 0.2\text{--}0.3$	Corrective action taken for important failure modes only.
$\alpha < 0.2$	Reliability has low priority.

efficiency of the growth programme. The value of α lies between 0 and 1, although 1 can never be achieved in practice.

O'Connor (1991) gives comments on the efficiency of a programme as a function of α, shown in Table 4.2. These give very good guidance, but it should be noted that higher values can be expected for an electronic equipment than for a mechanical one.

As was mentioned earlier, T/n is the cumulative MTBF, M_c, and is less than the MTBF that would be observed were the development programme to be stopped at this point, as the causes of failure have been fixed, and the failures observed up to time T should not recur, or at least should have a lower failure rate. The equipment's actual MTBF is known as the instantaneous MTBF, is denoted M_i, and the relationship between the instantaneous and cumulative MTBFs is:

$$M_i = \frac{M_c}{(1 - \alpha)}. \tag{4.2}$$

The graph of $\ln(M_i)$ is a line parallel to that of $\ln(M_c)$ but when drawn on log-log paper lies a distance $\ln(1 - \alpha)$ above it. The derivation of this formula is given in Section 4.2.4.

As an illustration, the data of Table 4.1 are analysed. Fig. 4.1 shows the Duane plot, drawn on log-log paper, and shows the plot of M_c, and above it the plot of M_i.

From the plot, α is 0.46. At the cumulative test time of 8245 hr, by the use of eqn 4.2, the instantaneous MTBF is given by

$$M_i = \frac{824}{(1 - 0.46)} \text{ hr}$$

$$= 1526 \text{ hr}.$$

Estimates of other parameters can be obtained from the plot, for example

(a) If the growth programme is continued until the equipment has been exposed to 10 000 hr of development time, what would the MTBF then be?

(b) The required MTBF is 2000 hr, how much more development is needed to reach this value?

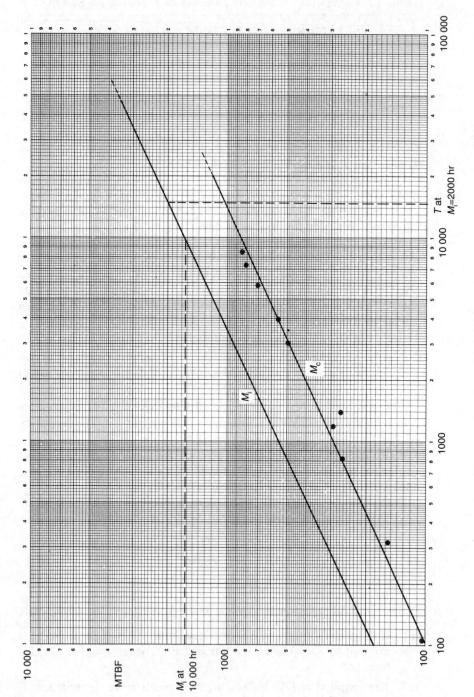

Fig. 4.1 Duane plot of the data of Table 4.1.

The estimates of these values can be obtained from the plot by observing where the extended M_i line takes the appropriate values of the parameters of interest, or in other words, extrapolate.

(a) The line crosses the point at which $T = 10\,000$ hr when M_i is 650 hr.

(b) The line crosses the point at which $M_i = 2000$ hr when T is 14 500 hr.

The question asked how much more time would be needed, and as this is the total time needed, the answer is the difference, $14\,500 - 8245 = 6255$ hr.

4.2.3 Further discussion and examples

Like most techniques, Duane plots must be used with care. They do show, often in a qualitative way, where the project is having difficulties. Consider the plot shown in Fig. 4.2. (Note: the grid from the log-log paper is not shown on this, or subsequent, Duane plots, for clarity.) Although the points appear to show that M_i is generally increasing, there are regions where it appears to be decreasing. These may be due to a change in personnel, who have a slightly different attitude, or who have to go through a learning process, or it may indicate a change in policy (maybe fewer resources are being dedicated to the reliability programme), or it may indicate a reliability problem that took some time to solve. In a customer/contractor relationship, it may well be a point that the customer would want to investigate further with the contractor, in a design review meeting for example, or during a technical risk assessment.

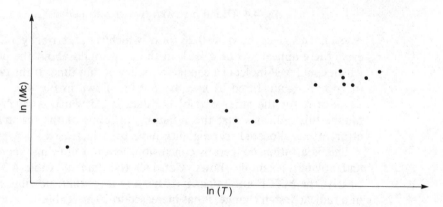

Fig. 4.2 Duane plot showing some growth.

The plot of Fig. 4.3 is interesting, and shows the line of best fit (fitted

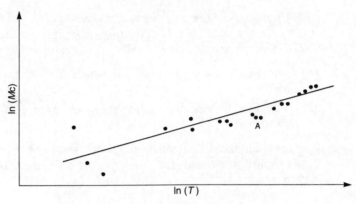

Fig. 4.3 Duane plot showing apparent growth.

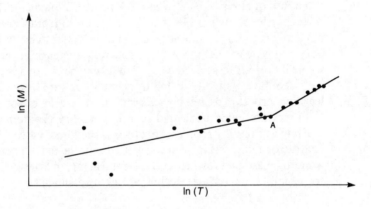

Fig. 4.4 Duane plot with two growth periods.

by eye). This gives the value 0.26 for α, which is not terribly good. However, there appears to be a kink in the data round about the point A, which could well indicate a change of policy at that time. If the two lines of best fit, again fitted by eye, are put in, shown in Fig. 4.4, then the value of α for the latter part of the data is 0.5 which is much better. Maybe this indicates that the situation, in terms of the resources and effort being allocated to reliability have been increased.

This is a rather dangerous conclusion, as the following, very simple and obvious, example shows. Consider the data of Table 4.3. Quite clearly, the MTBF until 40 hr is 10 hr, and after that, possibly because of a radical design change, it has increased to 20 hr. Table 4.3 also shows the values of M_c that would be obtained by blindly following the formula for a Duane plot. These data are now plotted, and this is done on Fig. 4.5, and shows just the sort of kink that was in Fig. 4.4, with a value of 0.34 for α for the latter part of the plot.

Table 4.3 Hypothetical growth data

Failure number (n)	Cumulative time (hrs)	M_c
1	10	10.0
2	20	10.0
3	30	10.0
4	40	10.0
5	60	12.0
6	80	13.3
7	100	14.3
8	120	15.0
9	140	15.6
10	160	16.0

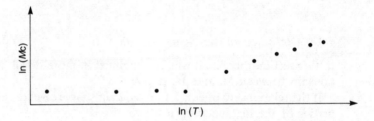

Fig. 4.5 Duane plot of data of Table 4.3.

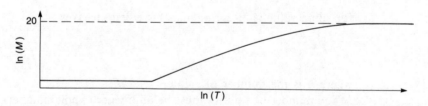

Fig. 4.6 Duane plot of data of Table 4.3: extrapolated into the future.

Of course, if it were possible to collect enough data, it would be clear that the value of M_c was approaching 20 hr in an asymptotic fashion, and the plot would be that shown in Fig. 4.6, but in a real situation, it would not be possible to collect that much data. Remember, too, that there is no statistical noise in this example, as there would be in a real situation. One way out of dealing with this situation is to replot just the data that were generated after the kink, and this is shown in Fig. 4.7 for the original problem, where the value of α is only 0.13, which is not nearly as good. Even this has its problems, but before we can go into this any further, it is necessary to explain more details of the model, and this is done in the following section.

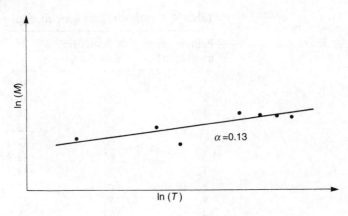

Fig. 4.7 Duane plot of the data used in Figs 4.3 and 4.4 taken after the Point A.

4.2.4 Derivation of the Duane equation

In this section the model will be explained in more detail, and the relationship between M_i and M_c is given.

If the relationship of eqn 4.1 holds, then simple algebra, using the properties of the log function, leads to

$$\frac{T}{n} = M_c = AT^\alpha \qquad (4.3)$$

for some value A. Rearranging this equation gives

$$n = \frac{T^{1-\alpha}}{A} \qquad (4.4)$$

where n is the number of failures.

The number of failures must be an integer (whole number), and this equation gives the relation between the *expected*, or *average*, number of failures. In theory, there is a probability distribution, for each time value, of the number of failures that could have occurred by that time. What we are concerned with here is the average of each of these distributions, at each time.

The *instantaneous failure rate*, λ_i, is failures over time, or the number of failures divided by the time, or in this case, as the failure rate is changing with time, the derivative of the expected number of failures with respect to time, dn/dT. This gives

$$\lambda_i = \frac{dn}{dT} = (1 - \alpha)\frac{T^{-\alpha}}{A}$$

so that the instantaneous MTBF, M_i, is given by:

$$M_i = \frac{1}{\lambda_i} = \frac{AT^\alpha}{1-\alpha}$$

$$= \frac{M_c}{1-\alpha}$$

which, substituting from eqn 4.3 is eqn 4.2.

The difference between M_i and M_c is illustrated in Fig. 4.8, where the curve shows how n varies with T, given by eqn 4.4. The slope of the chord is λ_c and its reciprocal is M_c and the slope of the tangent is λ_i and its reciprocal is M_i.

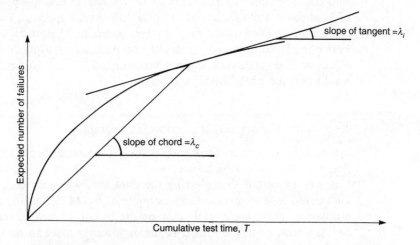

Fig. 4.8 The number of failures as a function of cumulative test time.

Note that the model assumes that any value of M_i can be achieved if development goes for a sufficiently long time, but that the rate of increase of M_i decreases, so that it eventually is not cost effective. This is illustrated by the following example. Table 4.4 shows the development time needed to reach the reliability target shown at the top of the table,

Table 4.4 Amount of testing needed to achieve a given target during a reliability growth trial

Current equipment reliability	Reliability target				
	50%	60%	70%	80%	90%
2%	175	340	705	1720	7830
6%	95	185	380	930	4225
19%	30	65	130	315	1440
30%	15	30	65	155	700

given that the equipment had already attained the value shown on the left hand side of the table. This is for a cross country vehicle that is subject to a lot of use, and the time is measured in weeks.

Suppose that the current equipment reliability is 19%, then to reach 60% would require 65 weeks. If there were 3 prototypes, which is generally considered to be the minimum acceptable, this is just over 20 weeks on each, or about 5 months. The time taken to do the fixes must be included in the equation, so we are talking about something like 2 years of real time for the development.

If 70% were required instead of the 60% discussed above, then the development time is 130 weeks, or about double (and so are the costs!). This could be achieved by having six prototypes, or taking 4 years instead of 2, or some combination of these. If 80% were required, then the time more than doubled again, maybe to ten years, or 15 prototypes. If 90% were considered necessary, then the resources are multiplied by a factor of about five, yet again! This is becoming silly, and targets like these would become prohibitively expensive.

4.2.5 Further considerations of the last example

It is now possible to go back and reconsider the example of Section 4.2.3 that was shown in Figs 4.3 and 4.4.

It was suggested that plotting the data that was generated after the kink could lead to a reasonable estimate of the MTBF at the time. The problem is that Duane plots assume the MTBF is zero at the start of development, or at the time that the data were started to be generated, and this is clearly not the case in this situation. One solution is to assume that the data are being generated after development time t_0 with n_0 failures already observed. This means that instead of plotting $\ln(t/n)$ as a function of $\ln(t)$ it is necessary to plot $\ln((t + t_0)/(n + n_0))$ as a function of $\ln(t + t_0)$. This is not easy, as t_0 and n_0 also have to be estimated. There are mathematical techniques that can deal with this, but they are not simple, and a computer is needed to do the number-crunching. There is certainly not a simple graphical technique, which is one of the advantages of Duane plots. It may be possible to use engineering judgment to estimate these parameters, but if that is done it could be difficult to justify the values chosen.

One approach is to plot n against t. This is shown for the problem in hand in Fig. 4.9. According to the previous section, these points should lie approximately on the curve shown in Fig. 4.8, and the failure rate is the tangent to this curve. In this case, the points appear to lie more or less on a straight line, or rather one of two straight lines, one for the first section of the data and the second for the latter section, where the point B in Fig. 4.9 corresponds to the point A in Figs 4.3 and 4.4. This

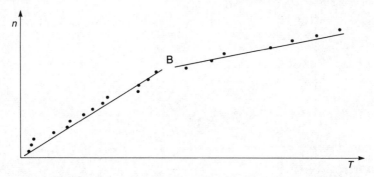

Fig. 4.9 The number of failures as a function of cumulative test time for the data used in Figs 4.3 and 4.4.

would indicate that there was little growth in the first stages of the programme, and then an effort was made to improve the reliability at the point A in Figs 4.3 and 4.4, or point B of Fig. 4.9 after which there was again little growth. The slopes of the two lines are estimates of the failure rates during these phases.

This example is based on real data that was analysed by the author, and the company concerned, who, when the kink was pointed out to them, admitted that they had made a large effort at that point. The author also tried the technique of crediting the programme with some time and failures, as described above, but the values estimated were so large that they swamped the data. This would have meant that the part of the curve of interest was so far to the right in Fig. 4.8 that reliability was growing very slowly anyway—which was the conclusion drawn from plotting n against t.

A further interesting example is shown by the n–t plot shown in Fig. 4.10. There are two very nice periods of growth—curves that are very steep initially, and then start to flatten out as the reliability grows. It is just a pity about the way the reliability suddenly worsened in the middle. In practice, this was due to design changes that were put in to improve other performance parameters. If it is necessary to obtain an estimate of the MTBF of the system at the time T, then it is quite reasonable to do a Duane plot of the data generated after all these changes: that is after A. This is shown in Fig. 4.11, and the value of α is 0.41.

The lesson to be learnt from these examples is that for any model, do not apply the model blindly, but use some judgment. It is always a good idea to plot the number of failures against time—in broad terms, if the result looks like a curve that is concave downwards, there is growth there, while if it looks like a straight line there is no growth, or little that can be measured!

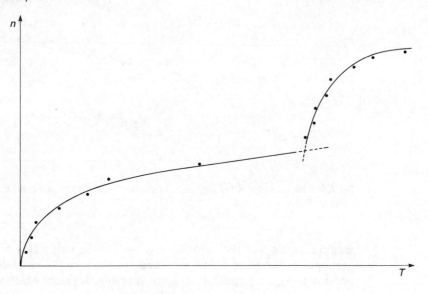

Fig. 4.10 Further example of number of failures as a function of cumulative test time.

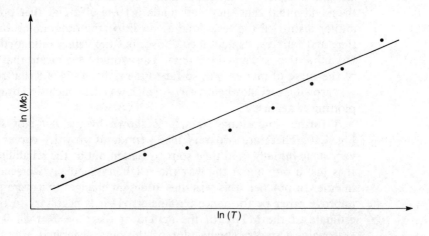

Fig. 4.11 Duane plot of the data of Fig. 4.10 taken after Point A.

In the case that the curve is concave upwards, then the reliability is getting worse, and no further comment will be offered here concerning that situation!

The Duane model does have its shortcomings, among which are that it assumes:

that the failure rate varies continuously with time;

that the fixes are done as soon as the failures occur;

that the failure rate is zero when time is zero, and can grow indefinitely;

no physical process, but is just a curve fitting exercise as a result of Duane's observation.

Despite this, it is very commonly used, mostly because of its simplicity and ease of application. It also puts pressure on contractors or management. There are other models, however, that do try to address some of these problems. None of them deal with all the problems, and in particular, none of them can deal with the last point.

When having to make decisions as to how many resources to put into a programme some analysis and prediction has to be done, and all predictions have to be considered on their merits. In a contractual situation, however, if a decision has to be made concerning the length and/or cost of a development programme, the above comments must be remembered. In the final analysis, the success or failure of a reliability growth programme depends on the skill and experience of the engineers concerned, and the commitment of all involved, engineers, management, and the customer.

For further comments on this technique, O'Connor, (1991), and Carter, (1986). BS 5760 Parts 2 and 3 discuss Duane, and Part 6, *Guide to programmes for reliability growth*, has a section on it. The UK Defence Standards, 00–40 and 00–41 give some guidance, while the American defence standard, US MIL-DTD 781, *Reliability testing for equipment development, qualification and production* is very useful.

Exercises 4.1

4.1.1 Draw the Duane plot of the following data:

Failure number	1	2	3	4	5	6	7	8	9	10
Cumulative test time (hr)	50	51	120	150	250	290	400	480	630	830

Estimate α, and M_i at 830 hr. If the test is continued until a total cumulative time of 2000 hr has elapsed, what is the predicted estimate of M_i at that point? About how much more time is required if the system MTBF is required to be 300 hr?

4.1.2 Draw the Duane plot of the following data. Comment on the plot. Draw the linear plot of n as a function of T. Redraw the Duane plot using an appropriate subset of the data. What is the value of α? Estimate the system MTBF at 1960 hr.

n	1	2	3	4	5	6	7
t	40	52	102	220	340	780	1010

n	8	9	10	11	12	13	14
T	1080	1170	1200	1280	1440	1560	1960

4.2.6 Fix effectiveness factors

Suppose that a failure is observed during development, and that this failure is analysed and, hopefully, fixed. The engineer cannot be certain that the cause of failure has actually been cured, so assigns a probability p to the chance of the failure being fixed. This probability is called the Fix Effectiveness Factor (FEF). When it now comes to estimating the MTBF, when the test time is divided by the number of failures, the reasoning goes that as there is now not one failure, but only $(1 - p)$ of a failure (on average), only $(1 - p)$ will appear in the denominator of the equation for the estimate of the MTBF. So if there are n failures in time T, and for the ith failure there is a FEF $p_i, i = 1, \ldots n$, then:

$$\text{MTBF} = \frac{T}{\sum (1 - p_i)} .$$

For example, reanalysing the data given in Table 4.1, reproduced in Table 4.5 with the FEFs and the running estimate of the MTBF after the fixes, the estimate of 2498 hr is obtained as the estimate of the MTBF.

Table 4.5 Fix effectiveness factors

Failure number (n)	Cumulative time (hrs)	M_c	FEF	MTBF
1	103	103	0.9	1030
2	315	157	0.8	1050
3	801	267	0.9	2002
4	1183	296	0.0	845
5	1345	269	0.7	791
6	2957	493	0.7	1478
7	3909	558	0.9	1861
8	5702	713	0.8	2479
9	7261	807	0.4	2503
10	8245	824	0.6	2498

The reader will notice that:

The cause of some failures is never found, and so they cannot be fixed, in which case the FEF is zero, as with failure number 4 in the table.

Duane modelling is like a FEF model with an average FEF of α.

This approach appeals to many design engineers, as it allows engineering judgment to be input into the analysis. It does avoid the difficulties of the Duane model, the criticisms of which can be applied to most analytic models, but it is not without its own problems, some of which are:

- It is a point estimate, and classical statistics cannot put confidence limits on this estimate. In Chapter 6 we see how Bayesian statistics can get round this problem, but at the cost of added complexity.

- It does not give any predictions regarding development time etc.

- The FEFs are subjective, and can be difficult to justify.

4.3 Acceptance trials and quality assurance

4.3.1 Introduction

Acceptance Trialling, Quality Assurance (QA), or Reliability Demonstration Trials (RDT) is trialling equipment with a view to making a decision as to its acceptability. It is not uncommon for a customer to demand an RDT at the end of development in order to demonstrate the equipment's reliability along the lines of '... the contractor will demonstrate that the equipment reliability is at least $X\%$ with a confidence of $Y\%$.', where Y is typically 95. This approach is erroneous, and shows confusion on the part of those particular customers, who are confusing confidence levels with risk. Consideration of the confidence levels may be appropriate when analysing data in other situations, but the intelligent customer must understand the risks that he is taking, and the risks the contractor is taking, when a demonstration test is agreed upon.

The problems with an RDT are:

- They are not actually a demonstration trial, rather an assurance trial; the reliability of equipment cannot be known until it has been in service for a long time, and often not until it is decommissioned, or in the case of a one shot device, used (when it is too late).

- Unlike growth trials, they do not contribute directly to reliability, in that if a fault manifests itself, there is no provision for fixing it – the trial is on the finished product, as it will go into service.

- It is difficult if not impossible to collect enough data to be sufficiently significant.

- The trials are frequently not carried out in an environment and under conditions that the equipment will meet in use.

- There is a problem about what to do if the equipment fails the test, as both parties are very keen for it to pass.

This does not mean, however, that RDTs have no part in the development of reliable equipment. As long as both parties fully understand the risks that they are running, and with adequate safeguards, they can play a significant role in helping to ensure equipment reliability. They are rather like the examination at the end of a course. Exams do nothing for the candidate's education, but they do ensure that he or she does some work. If nothing else, for a contractor to know he has to submit his finished product to an RDT does help him to concentrate the mind.

4.3.2 Initial ideas

The idea is best illustrated by an example. Suppose a customer is considering purchasing a large batch of simple items. He tests a small sample, and may, for example, test 50 items, and if none of them fail or if just one of them fails to function properly, then he accepts the batch as being of acceptable quality, but if two or more fail then he rejects it as being of unacceptable quality. (Statisticians are simple people, and divide the batches into two categories, those that are accepted and those that are rejected. In practice, a rejected batch may be accepted at a concessionary price and used for training, or some other decision may be made apart from the simple one of sending it back.) Such a procedure is called a sampling scheme.

If p is the proportion of defective items in the batch (i.e. the proportion of items that will fail to function when required to do so), called the failure ratio, and the batch is so large that p does not vary when some of the items are removed and tested, then the probability of accepting the batch, P, can be obtained using the binomial distribution, and is

$$P = P(0 \text{ or } 1 \text{ defectives in } 50)$$
$$= q^{50} + 50\,pq^{49}.$$

The graph of this equation is shown in Fig. 4.12. The formula is called the Operating Characteristic (OC), and its graph the OC curve.

Two important points on the OC curve are shown in Fig. 4.12. These are the values of p corresponding to the values 5% and 95% on the P(acceptance) axis, and take the values $p_1 = 9.13\%$ and $p_0 = 0.712\%$ respectively: i.e.

$$P = 5\% \text{ when } p_1 = 9.13\%$$

and

$$P = 95\% \text{ when } p_0 = 0.712\%.$$

In this case, the value p_1 of p is called the Lower Quality Level, or LQL, and is a measure of the consumer's risk, in that it is a poor quality at which there is still a significant probability of acceptance. This probability of acceptance is denoted β. The value p_0 is called the Upper Quality Level, or UQL, and is a measure of the producer's risk, being a good quality at which there is a significant probability of rejection. This probability of rejection is usually denoted α. The values of α and β may not be 5% or even be equal, but the values of α, β, p_0 and p_1 must be agreed by both sides.

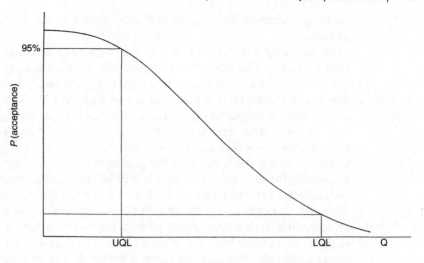

Fig. 4.12 Operating characteristic curve.

In the past the UQL has often been the quality the customer believes is suitable for his purpose, the idea behind this being that if he asks for a given quality, and contracts for it, the contractor expects to have a very good chance of him accepting the goods if he produces them at the required quality. The LQL on the other hand has been the worst quality that the consumer considers to be just acceptable to him. These days, however, it is more and more frequently the case that the specified reliability is, if not actually the LQL, a value that is very close to it, current thinking being that if a certain reliability is contracted for, there should be little or no risk of accepting something worse.

The ratio

$$D = p_1/p_0$$

is called the discrimination ratio, and is often used as a measure of the power of the test. The closer D is to unity, the steeper the OC curve, and the more powerful the scheme. Ideally the OC curve would be square, with a 100% probability of acceptance above the acceptable quality and a zero probability below it. For the scheme discussed above,

$$D = 9.13/0.712 = 12.8.$$

In order to improve the scheme, it is necessary to take a larger sample.

It is common for acceptance schemes to be chosen on this basis, that the consumer and contractor state the levels of risk that they consider acceptable, i.e. p_1 and β, and p_0 and α, respectively, and choose the scheme that most closely matches the requirements. With two points on the curve it is not a difficult matter (though a tedious one) to calculate the two parameters (sample size and acceptance number) that determine the scheme.

All RDTs have an OC, a UQL, and an LQL, and it is important that both contractor and customer recognise this and are prepared to accept the risks involved. In practice, it is very often too costly to run a trial that would reduce the risks to both parties to a very low level. As long as both parties appreciate this fact, however, they can manage the risk appropriately, and there is no reason why an RDT should not be run at a milestone as a decision aid, with the equipment being accepted into service should it pass, but with the contractor still accepting some responsibility for the equipment once it is in service, until enough data has accumulated for both parties to have an acceptable estimate of the reliability. It is also rarely the case that a project reaches the stage when an acceptance trial is appropriate without the two sides having some idea of the system's reliability. The acceptance trial is, in reality, to gain additional assurance. Bayesian statistics, Chapter 6, can be used to incorporate prior data and beliefs into the acceptance trial.

Example 4.1 A simple acceptance test for an MTBF may consist of testing for a fixed time and accepting only if the number of failures is below a stated number. So, for example, a customer may decide to test for 100 hr, and accept only if there are two or fewer failures. In that case, if a constant failure rate is assumed, the OC is derived from the Poisson distribution, and

$$P(\text{acceptance}) = P(2 \text{ or fewer failures})$$

$$= e^{-m} + me^{-m} + \frac{m^2 e^{-m}}{2!}$$

where m is the average number of failures expected in the 100 hr period, and is equal to 100/MTBF. Then if α is 5%

$$m = 0.818$$

and if β is 5%

$$m = 6.29$$

so the UQL and LQL are 0.818 and 6.29 respectively. These correspond to the values of 122 hr and 16 hr respectively for the MTBF, and give an indication of the respective risks the producer and consumer are taking. The range of these figures, from 16 to 122 hr, would indicate that the test is probably too slack, and should be tightened up. This can only be done by collecting more data, an expensive and time consuming task, although some of the test procedures described below can reduce the test times.

In the real world, trials may be more complex than the simple examples shown, and the conduct of the later trials may depend upon earlier results. Some assumptions may have to be made in order to calculate the

OC, LQL, and UQL, or it may only be possible to calculate limits on the risks for a range of assumptions. Whatever the situation, customer and contractor should be aware of the position when they come to an agreement on the conduct of the trial. Some examples are given in Section 4.3.5.

In summary, when designing an acceptance test, the following points should be considered.

(a) The conduct of the test must be decided very early in the project, and if it is a development project, must be written into the contract.

(b) When deciding on the conduct of the test, both customer and producer or contractor must be aware of the risks they are taking. For many simple situations there are standards describing the tests and the levels of risk.

(c) If the customer is purchasing a simple system such as ammunition, an assurance test may be appropriate, but when a complex and expensive equipment is being purchased, possibly after a long development programme, an assurance test may be too expensive to conduct, particularly if a very high reliability is required (in the case of safety equipment, for example). In this case, assurance can also be obtained from data generated during the development programme, or from historical data, particularly if similar equipments have been in use for some time, or from a knowledge of the company's record.

4.3.3 Some sampling schemes

MIL-STD 105D and BS 6001 were written with a view to testing one shot devices that either function or fail. This is called testing by attributes; i.e. items are examined one by one, and each is either accepted or rejected; i.e. each individual trial is either a success or a failure, the item either functions or it does not. There is not a parameter such as MTBF or MTTF being measured. The standards give schemes for a variety of UQLs and LQLs. This is done by assuming there is a level of quality acceptable to the customer, the Acceptable Quality Level (AQL) that is attributable to each scheme. The AQL is not derived mathematically. Given the AQL and the batch size, the appropriate plan can be determined.

In practice, the standard was developed to assure the quality of ammunition and other one-shot devices (flares, fire extinguishers, electric fuses, etc.) that were being purchased in large batches, but can be used for any situation in which an individual trial is considered either a success or a failure. The parameters are worked out using the binomial distribution, which is fine if the batch is large (which is the case in most instances) and in that situation the batch size does not matter. Intuition, however, suggests (wrongly) that the batch size should be taken into account even when the batch is large, and this intuitive though mistaken belief is pandered to in the standards.

As the standard was actually written to help with the assurance of the quality of manufactured items, and not the reliability of a design, parts of it are irrelevant to RDTs, but the mathematics and UQL and LQL are equally valid in both cases. The standard can be used to find a suitable scheme if the LQL and UQL required are known.

So for example, for a batch size of 281 to 500, if the AQL is 4% (the maximum proportion of defectives considered acceptable), then the standard proposes a sample of 50, with an acceptance number of 5 and a rejection number of 6 (accept if there are 5 or fewer defectives, and reject if there are 6 or more). The LQL and UQL corresponding to values of α and β of 5% are also given in the table, and are 19.9% and 5.34% respectively. The probability of acceptance if the contractor presents his goods at 4% defectives is 98.6%.

The standard also gives a double sampling scheme. This is a progressive scheme which attempts to reduce the cost of testing by keeping the sample size down in the following way. An initial sample of only 32 items is tested, and if there are 2 or fewer failures the batch is accepted, while if there are 5 or more the batch is rejected outright. If the number of failures falls in the grey area of 3 or 4 failures, then a second sample of 32 is tested, and if the total number of failures is 6 or fewer, then the batch is accepted, while if there are 7 or more, the batch is rejected.

The advantage of such a scheme is that if the goods are supplied at a reasonable level (near the AQL), then the chances of acceptance straight away is very high, and the second sample is probably not needed, thus reducing the costs of the trial. If the quality or reliability is very poor, it is also found out very quickly, and only one sample is needed. This is summed up in Fig. 4.13, which shows the average, or expected, number of items that have to be tested for a given presented quality. In this case the schemes are designed in such a way that the OCs are very close (the standard does not give the LQL and UQL for double schemes), as is shown in Fig. 4.14. It is easy to calculate, however, that the probability of acceptance for a quality of 4% defectives, the AQL, is 98.4% (cf. 98.6% for the single scheme).

This principle, of taking a number of samples until a decision is reached can be taken to further extremes, and the standard describes a seven fold scheme, called multiple sampling. In this up to seven samples can be examined, at each stage a definite decision regarding the acceptability or otherwise of the batch or design is made, or the evidence is not sufficiently clear cut, and a further sample is taken and examined. Table 4.6 shows the parameters of a multiple sampling scheme with very similar parameters to the schemes discussed in the previous paragraphs.

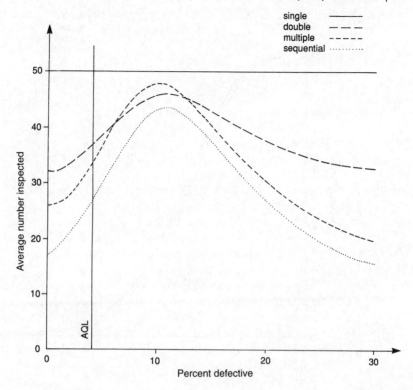

Fig. 4.13 Comparison of average numbers inspected for different types of inspection schemes.

The # sign indicates that a decision to accept cannot be made at that stage. So from Table 4.6, if there are fewer than four defectives in the first sample of 13, a second sample is taken, otherwise the batch is rejected. If there are one or no defectives in total after examining the second sample of 13, then the batch is accepted, while if there are a total of five or more defectives, the batch is rejected. If there is a total of two, three or four defectives, then a third sample of 13 is taken, and so on. There may be as many as seven samples taken, in which case, if there is a total of nine or fewer defectives in the total of 91 items examined (seven samples of 13 items each), then the batch is accepted, otherwise it is rejected.

The parameters of this scheme are similar to those of the earlier schemes, as was said above. The OC curve is drawn on Fig. 4.14 with the OC curves of the previous schemes, and it can be seen that the differences are very small. The saving in effort, as measured by the average number of items examined, is illustrated in Fig. 4.13, where it can be seen that the average is generally less than that for the double sampling

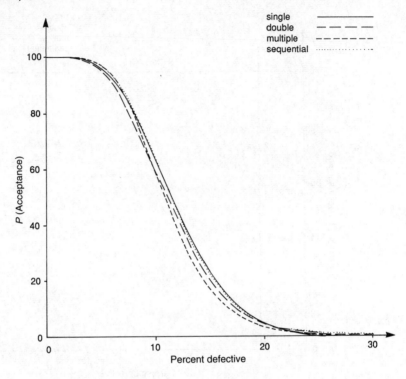

Fig. 4.14 Comparison of OC curves for different types of inspection schemes.

Table 4.6 Example of multiple sampling scheme

Cumulative sample	13	26	39	52	65	78	91
Acceptance no.	#	1	2	3	5	7	9
Rejection no.	4	5	6	7	8	9	10

scheme, and in particular, it is less when the quality or reliability is round about, or better than the AQL.

4.3.4 Sequential sampling

Sequential sampling is of a different nature altogether, and when testing by attributes, the conduct of the test is as follows. As items are examined, they are scored according to whether or not they are found acceptable, a positive score for items that pass and a negative score for items that fail. The cumulative score is recorded, and the batch is accepted if the score becomes sufficiently high at any stage, and rejected if it ever falls below some lower limit. In theory the sampling can go on indefinitely, but in practice a maximum sample is set.

The way this works in detail is that initially the cumulative score is set at a value called the handicap, H. One is added for each item that functions properly, and a value called the penalty, b, is deducted for each item that fails. As soon as the cumulative score reaches the target, which is $2H$, the batch is accepted, while if the score becomes zero or negative the batch is rejected. In theory the sampling could continue indefinitely, but in practice it is truncated when the number of items examined reaches the total sample size of the appropriate multiple sampling scheme, and if this number of items has to be examined, then the multiple sampling plan accept and reject numbers are used to decide on the fate of the batch or design.

If n is the total number of items examined at some point, and there have been r failures, then the score is given by

$$H - rb + (n - r).$$

The rules state reject if this is zero or negative, i.e. if

$$H - rb + (n - r) \leqslant 0$$

which on rearranging gives reject if

$$r \geqslant \frac{H + n}{b + 1}$$

and accept if the score reaches $2H$, i.e. if

$$H - rb + (n - r) \geqslant 2H$$

which on rearranging gives accept if

$$r \leqslant \frac{n - H}{b + 1}.$$

This divides the n,r plane into three regions, accept, reject, and continue testing, as shown in Fig. 4.15. Figure 4.15 also shows the effect of truncation.

The scheme that corresponds to those discussed earlier in this section, for example, has a handicap of 17 and a penalty of 8. Table 4.7 shows a log of such a scheme in practice. The OC curve of this scheme is also shown in Fig. 4.14, and it can be seen that it agrees very closely with the other schemes discussed earlier, while Fig. 4.13 shows how much more economic it is, on average, than the others.

Sequential sampling schemes for testing by attributes can be found in BS 6001 and BS 5760 Part 10, Section 10.5. Further discussion can be found in O'Connor (1991) and Grosh (1989).

This idea can be extended to testing an MTBF. In this case, the number of failures, n, is plotted as a function of the total time on test, T, and the n–T plane divided into the three regions, accept, reject, and continue testing. Such a scheme is illustrated in Fig. 4.16, along with a staircase

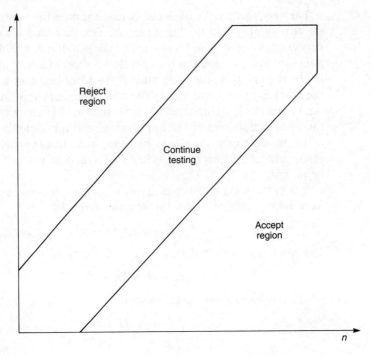

Fig. 4.15 Accept-reject regions in the *n–r* plane for sequential sampling.

plot showing the history of a trial. The time on test is the cumulative time, and may be on a number of components or systems. It is assumed the failure rate is constant.

The American defence standards US MIL-STD 781 and US MIL-HDBK 781 give extensive descriptions of the method and give test plans. They were originally developed for testing electronic components, but are equally applicable to systems, as long as the constant failure rate assumption is acceptable.

Exercises 4.2

4.2.1 A manufacturer of fireworks tests the reliability of his product by firing off 200 fireworks, chosen at random, each week. If only one fails, this is considered acceptable, but if two or more fail, some action is taken. What are the UQL and LQL, if α and β are each 5%?

4.2.2 A prototype of an equipment that is to work in a harsh environment is put on trial prior to going to production. It is decided to test it for 1000 hr, and if there are four or more failures, delay production while the reliability is improved. What are the UQL and LQL, if α and β are both 5%?

Table 4.7 Examples of sequential sampling

Item no.	Example 1		Example 2		Example 3	
	Pass	Score	Pass	Score	Pass	Score
1	Yes	18	Yes	18	No	9
2	Yes	19	Yes	19	Yes	10
3	Yes	20	No	11	Yes	11
4	Yes	21	Yes	12	No	3
5	Yes	22	Yes	13	Yes	4
6	Yes	23	Yes	14	No	−4
7	Yes	24	Yes	15	Batch rejected	
8	Yes	25	Yes	16		
9	Yes	26	No	8		
10	Yes	27	Yes	9		
11	Yes	28	Yes	10		
12	Yes	29	Yes	11		
13	Yes	30	Yes	12		
14	Yes	31	No	4		
15	Yes	32	Yes	5		
16	Yes	33	Yes	6		
17	Yes	34	Yes	7		
18	Batch accepted		Yes	8		
19			No	0		
			Batch rejected			

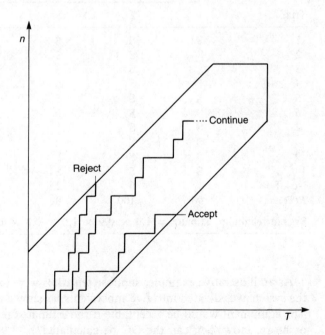

Fig. 4.16 Accept–reject region in the n–T plane for sequential sampling for an MTTF.

4.2.3 A piece of equipment is required to have a reliability that corresponds to an MTBF of 10 000 hr. A customer demands an RDT, when it is explained to him that this requires the accumulation of at least 50 000 hr of running time on the prototypes (over six years). As this is impracticable, how else might the contractor give the necessary assurance?

4.3.5 Some further examples

Demonstration tests do not have to go along the lines outlined in the previous sections, and in this section two examples are produced to illustrate just what may be done.

The first example consists of a system that had a fairly simple task, which could be tested, though at great expense. At the end of a trial, the system could be examined and the cause of any failures determined, in terms of the subsystem in which they occurred. As a result, at the end of a number of trials, the reliability of each of the subsystems could be estimated. A possible series of ten such trials is shown in Table 4.8. It was decided to run an acceptance trial along these lines, and if the estimated reliability was above a predetermined value, the equipment would be accepted by the customer.

Table 4.8 Example of an acceptance trial

	Subsystem				
Trial	1	2	3	4	5
1	S	S	S	S	S
2	S	S	S	S	S
3	S	S	S	S	S
4	F	S	S	S	S
5	S	S	S	S	S
6	S	S	F	S	S
7	S	S	S	S	S
8	S	S	S	S	S
9	S	S	F	S	S
10	S	S	S	S	S
No of Fs	1	0	2	0	0
$R(\%)$	90	100	80	100	100

System reliability estimate $= 1.0 \times 0.9 \times 1.0 \times 0.8 \times 1.0 \times 1.0 = 0.72$.

As an illustrative example, suppose ten trials were to be run, and that the system was divided into five subsystems, as shown in the illustration. The equipment would be acceptable if the estimated reliability was 80% or more. How then can the OC be calculated?

Suppose that the reliability of the ith subsystem is R_i, $i = 1, \ldots, 5$. The equipment reliability is given by the product of these, i.e:

$$R_S = \prod R_i.$$

Then if there are no failures, clearly the equipment will be accepted, and the probability of this happening is

$$P_0 = \prod R_i^{10}.$$

If there is just one failure, then the equipment will also be accepted, as the estimated reliability will be 90%. The probability of this happening is

$$P_1 = \sum_{j=1}^{5} 10R_j^9 F_j \prod_{i \neq j} R_i^{10}.$$

If there are two failures, then either they both occur in the same subsystem, in which case the estimated reliability is 80%, and the probability of this happening is

$$P_2 = \sum_{j=1}^{5} 45R_j^8 F_j^2 \prod_{i \neq j} R_i^{10}$$

or they occur in two distinct subsystems, in which case the estimated reliability is 81%, and the probability of this happening is

$$P_2' = \sum 90R_j^9 F_j R_k^9 F_k \prod_{i \neq j, k} R_i^{10} \qquad (4.5)$$

where the summation is over all possible values of j and k between one and five. (Note: this assumes that the two failures cannot occur in the same trial; i.e. it excludes the case that there may be two distinct, independent failures in one trial. If this case is not to be excluded then in eqn 4.5 the 90 just after the summation sign is replaced by 100.)

If there are three or more failures, then the system will not be accepted, so the probability of acceptance will be the sum of the above expressions. In order to calculate the OC, which must be a function of the system reliability, some assumptions must be made. One such could be that all the subsystems are equally reliable, i.e. that all the R_i take a common value R, with

$$R_s = R^5$$

in which case the OC is given by:

$$R^{50} + 10R^{49}F + 135R^{48}F^2.$$

Another approach might be to make some assumptions about the relative values of the R_i, such as that one of the subsystems was far less reliable than all the others, and that its failure probability is always twice the failure probabilities of the others, which are identical, or, what would probably make the calculations easier, that its reliability is the square of the other subsystem reliabilities.

Other assumptions which are far less easy to deal with concern the way in which the subsystems fail and the effect on the rest of the system. Possibly, certain failure modes in one or more of the subsystems prevent the rest of the system being exercised, in which case trials where that particular failure mode occurred would contribute no information about the rest of the system. This can be taken into account in the equations, but it makes the calculations more difficult and complex.

Another approach would be to put limits on the value of the OC. This is a fine exercise in multivariable differential calculus, to calculate the maximum and minimum values of the expression for the OC for a given value of the system reliability, but is rather long winded and neither the calculations nor the results will be given here. The bounds obtained are almost certainly very wide, and although of interest to the customer, who would be most interested in the worst case, while the contractor would be interested in the best case, are not really of much value unless some other assumptions are made as well.

The above description gives some of the difficulties involved in running a trial in this way. Apart from the mathematical difficulties, this type of trial is not very realistic. The customer is interested in the functioning of the system, and when it is in use, sees it as a 'black box', that either functions or fails, and the RDT should reflect that.

The second example is along very different lines, as the conduct of the trial included inducements to the contractor to ensure reliability. The customer was purchasing a relatively large number of the system under consideration, which would be put into store until needed. Owing to the constraints of manufacture etc., these were being supplied in small batches over a period of time, and it was possible for the customer to test each system. An acceptance plan was worked out, with a fixed UQL, giving acceptance criteria for each batch considered as a sample of the total number contracted for. The acceptance number was calculated on the cumulative sample size, using the hypergeometric distribution. Table 4.9 shows an example, where the total contract is for 100 items, and they arrive in 10 batches, or samples, of 10, with the contracted reliability, which is put equal to the UQL, at 90%, and α and β at 5%. The OC curves are shown in Fig. 4.17.

Note that the scheme gets tighter as time progresses, in the sense that if the contractor presents equipment which is just at the limit, the acceptance numbers for each sample either decrease or at best remain constant. This is a natural way for the contractor to acquire manufacturing experience or to go up a learning curve (if needed). Note too that at the UQL, the probability of acceptance must be at least 95%.

Table 4.9 Cumulative acceptance scheme

Sample number	Cumulative sample size	Cumulative acceptance no.	Probability of acceptance (%)
1	10	3	99
2	20	4	97
3	30	5	96
4	40	6	95
5	50	7	95
6	60	8	96
7	70	9	98
8	80	10	99.999
9	90	10	99.999
10	100	10	100

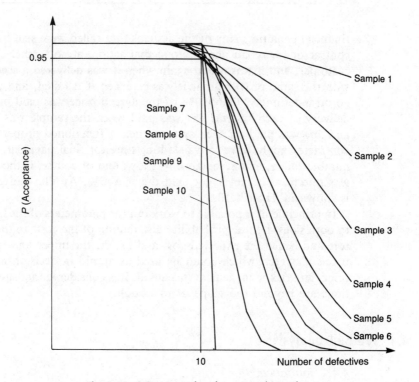

Fig. 4.17 OC curves for the example in the text.

Note that the OC curves get steeper and steeper as the scheme progresses, and that the final one is a step—as every item has been examined. Note too that if the samples just pass each time, the scheme gets tighter and tighter.

There was an added parameter to this scheme, in that there were

Table 4.10 Cumulative acceptance scheme example

| Sample number | Failures | | Remarks |
	In sample	Cumulative	
1	2	2	Accept
2	2	4	Accept
3	1	5	Accept
4	0	5	Accept
5	2	7	Accept
6	2	9	Reject, loss of 5% on Sample
7	1	10	Reject, loss of 5% on Sample
8	0	10	Accept
9	0	10	Accept
10	0	10	Accept

financial penalties. Any of the systems that failed were sent back to the contractor for repair and any remedial action, at no further cost to the customer, and it was tested again when it was delivered a second time, with the same proviso, that it was returned if it failed, and this could go on indefinitely. Only 90% of the agreed price was paid on the first delivery of each system, 5% was paid when the sample was accepted, and 5% was paid on each system when it functioned properly. So the contractor might have to wait a long time for final payment, if a large number failed, and acceptance was low, and of course, if the final trial was rejection, he never got paid the final 5%. An illustrative example is shown in Table 4.10.

In principle, it is possible to work out the parameters of the likely costs to both sides, i.e. the probability distribution of the costs to the contractor, and hence the expected cost and profit, the upper and lower 95th percentiles etc., which could be used as an aid in decision making on price, and risk management in general. It is considered that such calculations are beyond the scope of this book.

4.4 Weibull analysis

4.4.1 Introduction

When considering life data, there are often good reasons for believing that the item on test does not have a constant failure rate, particularly if an aging mechanism is involved, as with mechanical components. In this case, the next simplest case is when the failure rate is proportional to a power of time, i.e.

$$\lambda \propto t^k.$$

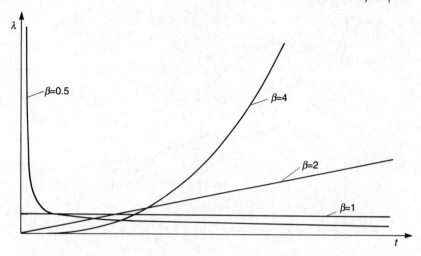

Fig. 4.18 Examples of the failure rate function for the two parameter Weibull distribution.

This gives rise to the Weibull distribution that was first presented in Section 2.5.7. The equations for λ, R, and f are presented again here, and they are, for the two parameter Weibull,

$$\lambda = \frac{\beta}{\eta} \left(\frac{t}{\eta}\right)^{\beta - 1}$$

$$R = \exp - \left(\frac{t}{\eta}\right)^{\beta}$$

$$f = \frac{\beta}{\eta} \left(\frac{t}{\eta}\right)^{\beta - 1} \exp - \left(\frac{t}{\eta}\right)^{\beta}$$

where β is called the shape parameter, and η the scale parameter, or characteristic life. Note that η has the same dimensions as t, while β is dimensionless. Increasing and decreasing failure rates can be modelled by appropriate choice of the shape parameter, as follows:

$\beta < 1$, decreasing failure rate;

$\beta = 1$, constant failure rate (exponential, η is the MTTF);

$\beta > 1$, increasing failure rate.

Graphs of λ, R and f are shown in Figs 4.18, 4.19, and 4.20 respectively.

This distribution, of which f above is the probability density function, is called the Weibull distribution, after the Swedish engineer, Weibull, who first used it and studied its properties. It is very often used in the

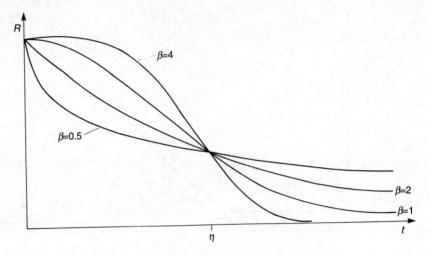

Fig. 4.19 Examples of the reliability function for the two parameter Weibull distribution.

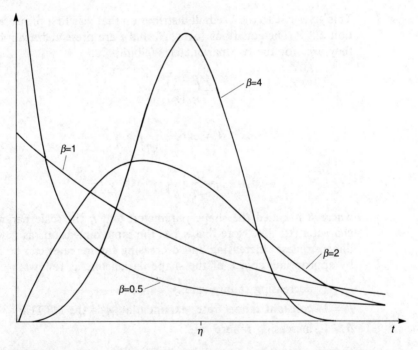

Fig. 4.20 Examples of the probability density function for the two parameter Weibull distribution.

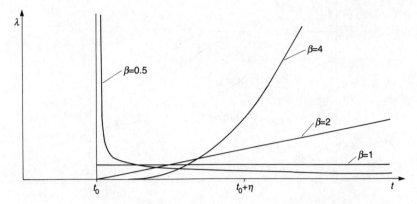

Fig. 4.21 Example of the failure rate function for the three parameter Weibull distribution.

analysis of life data, as there are good theoretical reasons for using it as well as the more heuristic ones like it is the next simplest model to the exponential, and it is easier to fit a two parameter distribution to a given data set than a one parameter one.

It is frequently extended further, by adding a third parameter, in which case it is known as the three parameter Weibull distribution. The equations for λ, f and R are then given by

$$\lambda = \frac{\beta}{\eta} \left(\frac{t - t_0}{\eta}\right)^{\beta - 1} \qquad t > t_0$$

$$= 0 \qquad t < t_0$$

$$R = \exp - \left(\frac{t - t_0}{\eta}\right)^{\beta} \qquad t > t_0$$

$$= 1 \qquad t < t_0$$

$$f = \frac{\beta}{\eta} \left(\frac{t - t_0}{\eta}\right)^{\beta - 1} \exp - \left(\frac{t - t_0}{\eta}\right)^{\beta} \qquad t < t_0$$

$$= 0 \qquad t < t_0.$$

The parameter t_0 is known as the minimum or guaranteed life.

Graphs showing λ, R, and f are shown in Figs 4.21, 4.22, and 4.23 respectively.

There is a simple graphical technique for estimating the parameters based on well known and documented statistical principles that is described in this section. The simplest case, where the failure rate is constant, is considered first as an illustration.

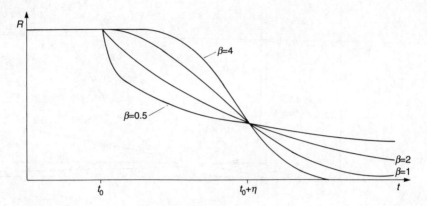

Fig. 4.22 Examples of the reliability function for the three parameter Weibull distribution.

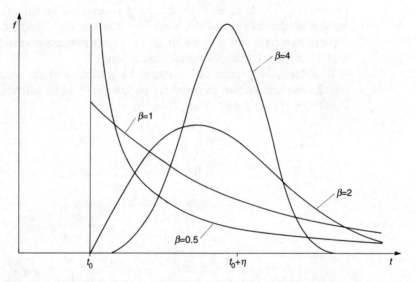

Fig. 4.23 Examples of the probability density function for the three parameter Weibull distribution.

4.4.2 Graphical estimation of a constant failure rate

If an engineer has data he believes comes from a system with constant failure rate, i.e. the data are exponentially distributed, the analysis is very simple. This data may consist not only of times to failure, but also censored data, that is to say times at which the system or components are still running. In this case, if T is the total test or running time, i.e.

the sum of the failure times and the censored times, and n failures have been observed, then the best estimate of λ, the failure rate, is given by

$$\lambda = n/T$$

or

$$\text{MTTF} = 1/\lambda$$
$$= T/n.$$

This is a classic statistical result and is called the Maximum Likelihood Estimate, or MLE, of λ and is discussed in Chapter 5. However, it fails to consider the possibility of the source of the data not having a constant failure rate, and the simple graphical goodness-of-fit test described below can deal with this difficulty illustrated by a simple example. Twenty components were tested for 80 hr, at which time the test was terminated. By this time, 8 of them had failed at the times shown in Table 4.11.

Table 4.11 Failure times of tested components

Failure number	1	2	3	4	5	6	7	8
Time (hr)	26	36	45	50	54	56	61	71

If the data come from an exponential distribution, then,

$$\text{MTTF} = 1359/8$$
$$= 170 \, \text{hr.}$$

Now consider the following analysis. In the case of the exponential distribution,

$$R(t) = e^{-\lambda t}$$

and taking natural logs, this gives

$$t = \frac{-\ln(R)}{\lambda}.$$

The values of t are obtained from the data. If the data are ordered, from least to greatest, so that t_1 is the time of the first failure, t_i the time of the ith failure up to $i = 8$ in the example, then plotting t_i against a suitable value of $\ln(R)$ should produce a straight line with slope $1/\lambda$, that passes through the origin. This technique gives a means of estimating λ as well as a graphical method of testing the goodness of the initial supposition that the data is from an exponential distribution. There is some discussion in the literature on the best values of R to be used. Statistical theory gives the means of calculating them by the use of order statistics. The following two expressions are the commonest used.

The median rank:

$$F_i = \frac{i - 0.3}{n + 0.4} \tag{4.6}$$

has the property that it will exceed the observed value half the time, and will be exceeded by it half the time, on average. This is opposed to the mean rank, which is the average value, and is given by:

$$F_i = \frac{i}{n + 1}. \tag{4.7}$$

An outline of the derivation of these formulae is given in Carter (1986). The advantages of the use of the median rank formula is that it is relatively easy to put confidence limits on the line, and that censored data can be dealt with. An example of the latter is given in Section 4.4.4.

An analysis of the data along these lines is shown in Fig. 4.24, which shows the data points and the line of slope 170 hr passing through the origin. The reader will observe that it is not a good fit, and that the points appear to lie on a curve rather than a straight line. It is possible that the data come from another distribution, such as the Weibull distribution, and the techniques for analysing Weibull data are discussed next.

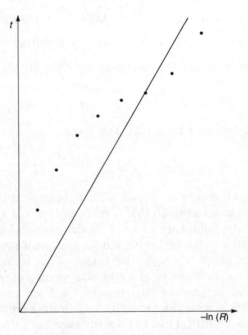

Fig. 4.24 Constant failure rate plot.

4.4.3 Graphical estimation of Weibull parameters

If the data come from a two parameter Weibull distribution, then

$$R(t) = \exp\left(-\left(\frac{t}{\eta}\right)^\beta\right)$$

take natural logarithms of both sides,

$$-\ln(R) = \left(\frac{t}{\eta}\right)^\beta$$

and taking logarithms again, and rearranging,

$$\ln(t) = \ln(\eta) + \frac{1}{\beta}\ln(-\ln(R)).$$

That is, a plot of $\ln(t)$ against $\ln(-\ln(R))$ will be a straight line with slope $1/\beta$ and intercept $\ln(\eta)$. In practice, it is not necessary to take the logs manually, as Weibull paper is available commercially, with a $\ln$ scale along the horizontal axis for t and a log-log scale along the vertical axis for F (not R). The use of an example of this paper is illustrated in Fig. 4.25, using the data of the previous example. The median rank values given in eqn 4.6 have been used for F, as before. Exact values for R have been calculated by White (1967) for use when doing Weibull analysis, as well as appropriate weighting factors if the line is fitted analytically using least squares.

An estimate of β, the shape parameter, is obtained by dropping the perpendicular from the estimation point in the top, left hand corner onto the line of best fit (estimated by eye in this case), and noticing where it crosses the scale marked $\hat{\beta}$, at the point marked B in the example, giving an estimate of β of 2.5.

The scale parameter, η, is estimated by observing that

$$R(\eta) = 1/e = 0.37$$

i.e. R is always 37%, and F always 63%, when $t = \eta$, and conversely t is η when $F = 0.63$, and so the value at which R is 0.37, and F is 0.63, is marked with a dotted line labelled η estimate. The point at which the η estimate line crosses the line of best fit, marked G in the example, gives the value of η, which in this case is 90 hr.

If the data is from a three parameter Weibull distribution, then applying the technique described above without taking t_0 into account will produce a curve rather than a straight line, as illustrated in Fig. 4.26. (Note: the grid from the Weibull paper is not shown on this or subsequent Weibull plots.) The procedure for estimating t_0 is as follows.

Pick three points F_i, $i = 1, 2, 3$, on the F axis with the properties that they are as far apart as possible, but still in the range of F that is covered

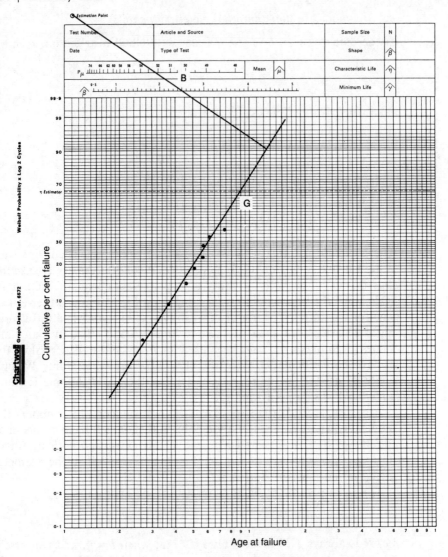

Fig. 4.25 Weibull plot.

by the data, and the distance measured on the Weibull paper between F_1 and F_2 is the same as that between F_2 and F_3, i.e.,

$$\ln(-\ln(R_2)) - \ln(-\ln(R_1)) = \ln(-\ln(R_3)) - \ln(-\ln(R_2)).$$

These correspond to three values of t, t_1, t_2, and t_3. The estimate of t_0 is given by

$$t_0 = t_2 - \frac{(t_3 - t_2)(t_2 - t_1)}{\cdot(t_3 - t_2) - (t_2 - t_1)}$$

$$= \frac{t_3 t_1 - t_2^2}{(t_3 - t_2) - (t_2 - t_1)}.$$

The derivation of this formula can be found in Carter (1986).

The value of t_0 so obtained can now be subtracted from each of the data points, and these new values plotted in the normal way, and it is from this second plot that β and η can be estimated. The data in Table 4.12 are analysed as an example. The initial Weibull plot is shown in Fig. 4.27, and from the curve the values

$$t_1 = 322$$

$$t_2 = 390$$

$$t_3 = 710$$

are obtained. This gives

$$t_0 = 304.$$

Table 4.13 shows the data with 304 subtracted from each reading, and this is plotted in Fig. 4.28, and is seen to be a reasonable straight line. The values of β and η from the plot are 1.2 and 187 respectively.

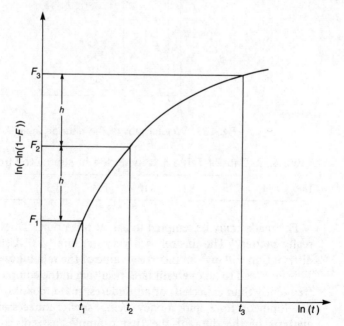

Fig. 4.26 Three parameter Weibull plot: theoretical curve.

Table 4.12 Failure data (hr) for three parameter Weibull analysis

322	358	379	396	414	458	507	520	585	710

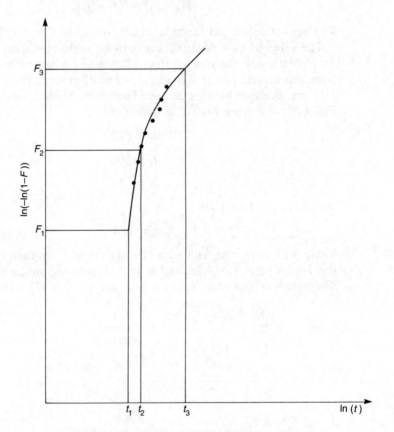

Fig. 4.27 Weibull plot of the data of Table 4.12.

Table 4.13 Data of Table 4.12 with 304 hr subtracted from each reading

18	54	75	92	110	154	203	216	281	406

The reader may be tempted to ask at this point, 'Does the distribution really matter?' The answer is a very definite '*yes*'. Using the incorrect distribution will give an incorrect value of the reliability at various times. This may lead to an overestimate, resulting in the equipment failing more frequently than expected, or an underestimate, possibly leading to more development time and money being spent unnecessarily. The initial analysis of the data of the first example, assuming it was from an exponential distribution, led to an MTTF of 170 hr, while the analysis

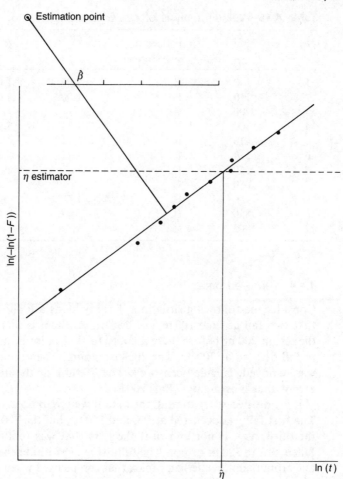

Fig. 4.28 Weibull plot of the data of Table 4.12 after subtracting t_0.

Table 4.14 Comparison of exponential and Weibull distributions

Time (hr)	10	20	50	100
Reliability (exponential)	0.94	0.89	0.75	0.55
Reliability (Weibull)	0.996	0.97	0.78	0.24

above concluded that the data were probably from a Weibull distribution with a shape parameter of 2.5 and scale parameter of 87 hr. Table 4.14 compares the values of the reliability at various times under these two assumptions.

Table 4.15 Weibull analysis of censored times

r	Time	Failure or Censored	I	R	F(%)
1	100	F	1.00	1.00	6.1
2	150	F	1.00	2.00	14.9
3	200	C			
4	300	F	1.11	3.11	24.6
5	410	F	1.11	4.22	34.4
6	520	F	1.11	5.33	44.1
7	560	C			
8	680	C			
9	750	F	1.67	7.00	58.8
10	850	F	1.67	8.67	73.4
11	1005	C			

4.4.4 Censored data

Consider the following situation. Twenty items are put on test, and the first two fail at time 100 hr and 150 hr. An item is then removed from the test at 200 hr before there are any further failures, and the next item to fail does so at 300 hr. The 200 hr reading, when the item in question was removed, is called censored. The lifetime of the item is unknown, except that it is greater than 200 hr.

It is now necessary to rank the data if we are to do a Weibull analysis. The first failures occurred at 100 and 150 hr, but the 300 hr failure is not the third, as it is not known if the item that was removed would have failed before 300 hr or not. The following algorithm generates the ranks, and from them the median ranks, that can be used when doing a Weibull plot.

The method is best illustrated by an example, and this is done in Table 4.15. The second column shows the times, T_r, both of failures and censoring, ordered, with the ordering, r, shown in Column 1. Column 3 shows if T_r is a failure, F, or a censored time, C. Column 4 shows the increment, I_r, which is obtained by use of the following iterative formula:

$$I_1 = 1 \text{ if the first time is a failure,}$$

$$= 0 \text{ if it is a censored time.}$$

$$I_r = I_{r-1} \text{ if } T_{r-1} \text{ is the time of a failure}$$

$$= \frac{N + 1 - R_s}{N + 1 - (r - 1)} \text{ if } T_{r-1} \text{ is a censored time,}$$

and R_s is the rank of the greatest failure T_s time such that

$$T_s < T_r.$$

Column 5 is the rank, R_r, which is given by

$$R_r = R_s + I_r \text{ if } T_{r-1} \text{ is a failure time}$$

$$= R_{r-1} \text{ if } T_r \text{ is a censored time,}$$

and

$$R_1 = 1 \text{ if } T_1 \text{ is a failure time}$$

$$= 0 \text{ if } T_1 \text{ is a censored time.}$$

Column 6 shows the failure probability, F_r, which is given by the formula of eqn 4.6. Note that I_r, R_r, and F_r are only defined for

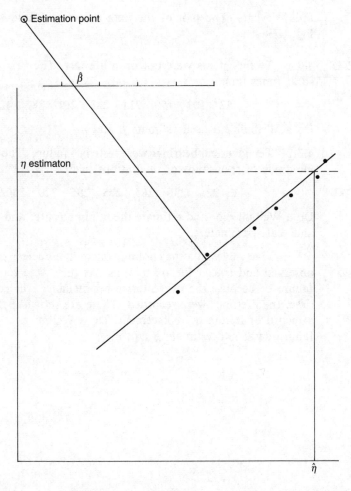

Fig. 4.29 Weibull plot of censored data.

Table 4.16 Valve failure and replacement data

Life (days)	Fail/replace	Life (days)	Fail/replace
25	F	71	F
89	R	48	F
101	F	110	R
33	F	35	F
58	R	98	R
29	F	68	F
103	R	72	F
52	F	80	R
59	F	100	F
65	R	31	F

failure events. The plot of the data shown in Table 4.15 is shown in Fig. 4.29.

Exercises 4.3

4.3.1 Twenty items were put on a life test. The first eight to fail did so at times (hr):

$$52 \quad 103 \quad 150 \quad 211 \quad 248 \quad 291 \quad 387 \quad 425$$

Do a Weibull plot and estimate β and η.

4.3.2 Ten identical bearings were tested to failure. The failure times (in hundreds of thousands of revolutions) were

$$211 \quad 221 \quad 236 \quad 265 \quad 275 \quad 295 \quad 320 \quad 340 \quad 400$$

Do a Weibull plot and estimate the minimum life, and hence the shape and scale parameters.

4.3.3 Five valves in a the cooling system in a nuclear power plant were observed and traced for six months. As they were replaced, either on failure or because the preventative maintenance was convenient at the time, the lifetimes were recorded. These are shown in Table 4.16, with a record of failure or replacement. Do a Weibull plot of the censored failure data, and estimate β and η.

Advanced statistics

5

5.1 Introduction

In Chapter 2 some statistical concepts were introduced, and descriptive statistics were examined in particular. In this chapter some more advanced concepts are examined. The ideas of sampling and data analysis, estimation, significance, and confidence are examined, paricularly when applied to a mean, and least squares fitting is described, and the ideas applied there.

5.2 Sampling and data analysis

5.2.1 Introduction

Suppose we have a number of lifetimes, obtained from observation, t_1, $t_2, \ldots, t_n$. If they are all more or less from the same source, i.e. there was nothing a priori that would make us believe that one would be any different from the others, then it is called a sample, from the population of all lifetimes of all the possible items in question. These lifetimes have a distribution, and it is required to obtain information about the distribution from the sample data—its mean, or standard deviation, for example.

Chapter 2 described how to deal with data in a descriptive fashion, how to group data to obtain a histogram, and how to calculate the mean and variance of a sample, giving the two formulae in eqns 2.2 and 2.3 for the variance. The use of these, and the reason for the two of them, will be described below. Just as the lifetimes themselves are distributed, so the sample mean and variance are distributed. That is to say, if the experiment were to be repeated, and more samples were examined, we would not expect them to have the same means or variances. As one might expect the sample parameter values to be more or less the same as the population parameter values, this variability must be dealt with. This is one of the objects of this chapter.

5.2.2 Estimation

An estimate of a population parameter is a function of the sample data which it is hoped is close in value to the true population value. So, for example, the mean of the sample mentioned above is an estimate of the population mean.

Statisticians spend a large amount of their time trying to find good estimates of parameters of interest. By good, there are some properties that are desirable in estimates. As an estimate is distributed, and it is possible, in theory, to find the distribution from a knowledge of the parent distribution of the sample population. The distribution of the estimate has an expected value and a variance. Ideally, the expected value of this distribution should be the true value of the parameter, in which case we say the estimate is unbiased, and the variance should be as small as possible, in which case the estimate is called minumum variance. It is a frequent, though not invariable, practice to use greek letters for the parameters of interest in the population, and either the corresponding Roman letter or Greek symbol with a hat (^) over it for an estimate or for the corresponding parameter of the sample.

When estimating a mean, the sample mean, $\bar{t}$, is the best estimate of the population mean (often denoted μ). (The $\bar{t}$ notation is the exception to the practice discussed in the previous paragraph.) It has the property that it is unbiased, and its variance is σ^2/n, where σ^2 is the population variance. It can be shown that $\bar{t}$ is approximately normally distributed, the approximation being better in the body of the distribution, i.e. near μ, gets better as the sample size increases, and is exactly normally distributed if t is normally distributed. Having said that, if t is exponentially distributed, which is the case of interest when estimating a constant failure rate, then $\bar{t}$ is Erlang.

When estimating a variance, an unbiased estimate of the variance of a distribution is given by eqn 2.3, which is reproduced here:

$$s^2 = \frac{\sum (t_i - \bar{t})^2}{N - 1}.$$

Note the $N - 1$, and not N, in the denominator. The similar expression with N in the denominator, given in eqn 2.2, is biased. The distribution of this estimate is not always easy to derive, but if the parent distribution is normal, the distribution of $(N - 1)s^2/\sigma^2$ is χ^2 with $N - 1$ degrees of freedom, and for a parent distribution that is approximately normal (i.e. reasonably bell shaped), the distribution of $(N - 1)s^2/\sigma^2$ is approximately χ^2.

A common estimate is the Maximum Likelihood Estimate, or the MLE. This is obtained in the following manner. Suppose we have some life data, $t_1, t_2, \ldots, t_n$ obtained by testing n items until they had all failed. Suppose further that the lifetimes came from a distribution with pdf $f(t)$. Then the probability of observing the ith lifetime, t_1, is given by P_i, where

$$P_i = f(t_i)\delta t$$

and the probability, P, of observing the sample of lifetimes, that is t_1 and t_2 and ... and t_n is given by the product of the P_i,

$$P = \prod P_i = \prod f(t_i)\delta t^n.$$

The product of the $f(t_i)$ in the above expression is known as the likelihood function, L, i.e.

$$L = \prod f(t_i). \tag{5.1}$$

As the t_i are known, L is a function of the parameters of the pdf $f(t)$, and the values of those parameters that maximizes L are the maximum likelihood estimates. They are functions of the t_i. The MLE is not necessarily unbiased or minimum variance, but nevertheless it is a useful starting point when seeking an estimate. The unbiased estimate of the variance discussed above, for example, is not in general the MLE.

Examples 5.1

5.1.1 Suppose the items on test are believed to have a constant failure rate λ, i.e. the lifetimes are believed to come from an exponential distribution. Then

$$f(t_i) = \lambda e^{-\lambda t_i}$$

and

$$L = \lambda^n \exp\left(-\lambda \sum t_i\right)$$

and it is a simple exercise in differential calculus to demonstrate that this function has a maximum when

$$\lambda = \frac{n}{\sum t_i}.$$

5.1.2 A sample of n identical items are put on test, and r of them fail, while $n - r$ function. Then if p is the probability of each item functioning, the probability, P, of observing this result is, from the binomial distribution,

$$P = \frac{n!}{(n-r)!r!} q^r p^{n-r}$$

where

$$q = 1 - p$$

as usual, and this is the likelihood function. It has a maximum when

$$p = \frac{n - r}{n}$$

and this is the MLE. It is also unbiased and has minimum variance.

5.3 Significance

5.3.1 Introductory example

Consider the following situation. Fifty items that are believed to be 98 per cent reliable are tested, and four of them are found to be faulty. What can be said about the claim that the reliablity is 98 per cent?

The reliablity looks as if it is only 92 per cent. However, the variability of this estimate must be taken into account. The estimate has a variability of its own, as the number of failures in a sample is binomially distributed. If the reliability is really 98 per cent, what are the chances of getting the result that was actually observed? If it is very small, then we must either believe that a very rare event has happened, or that the claim (that the reliability is 98 per cent) is untrue – i.e. we must reject it.

The probability of getting the result observed, or a worse one, is the probability of observing four or more defectives in the fifty and can be calculated from the binomial distribution, and turns out to be 0.02 or 2 per cent. So we must believe that an event has occurred that had a prior probability of occurrence of 2 per cent, or we reject the 98 per cent reliability claim.

In practice, a value for the limiting probability, α, is chosen in advance, called the significance level, and if the actual probability falls below this value α, we say the result is significant at that level. Values often chosen are either 5 per cent or 1 per cent, so we can say that the result in question is significant at the 5 per cent level (because our result is 2 per cent which is less than 5 per cent, and so an event that is less likely than 1 in 20 has happened), but that it is not significant at the 1 per cent level (because 2 per cent is greater than 1 per cent, and so the event that has occurred is not as unlikely as 1 in 100).

5.3.2 The conduct of a significance test

This section sets out just how significance tests are conducted in practice. Subsequently some worked examples are presented.

Initially a null hypothesis that is to be tested is chosen. (Significance testing is often called hypothesis testing.) This is the assumption that is to be tested—in the example above, it was that the reliability was 98 per cent. The null hypothesis, denoted H_0, is usually quantitative in nature, and may be of the form that the reliability is a value R, or that the MTBF is a value M, or it may be that they are at least those values, or that they lie within a particular range of values. In reliability work, H_0 is often that the MTBF or the reliability is at least a certain value. We are normally only interested in these parameters being too low, they cannot, in general, be too high. The null hypothesis is used in the calculation of the probability that tells us if the result is significant or not. The null hypothesis may also be of the form that the population distribution is of a particular type. These situations will also be examined later. An alternative hypothesis H_1, must also be chosen, and in the situations that are of interest to us, will be the converse of the null hypothesis.

The next step is to choose the significance level. This is a technical point, and often a number of levels are chosen for consideration, usually including 5 per cent and 1 per cent. There is often little logical reason for choosing one level rather than another, but some analysis may indicate favoured levels. These include a consideration of the possible errors that can be made. It may be that H_0 is true, but because of the statistical variability in the data, it is rejected as being false. This is called a type I error, and the probability of this happening is precisely the significance level chosen. On the other hand, it may be that H_0 is false, but that there is insufficient data to indicate this, in which case H_0 is accepted. This is called a type II error. It is impossible to quantify the probability of a type II error without making further assumptions, which may incude some subjective ones, but if it can be done, and if the costs of making a wrong decision are known, then the significance level can be chosen in a logical way.

For example, suppose that a decision has to be made regarding a new design of car. Development has reached a point where it must be decided whether to continue development for a further year, which adds to development costs, or to stop development and start production. If the reliability is sufficiently high, and the company goes for further development, then the money will be spent unnecessarily, while if the car is unreliable, and goes into production, and subsequently reaches the consumer, the latter will have more than the usual number of warranty claims etc., which will also be costly (to say nothing of the unquantifiable costs to the company's good name etc.). It is possible, in theory, to balance, these costs with a suitable choice of significance level, in such a way as to minimize the expected loss due to these costs. Note that this is a very simple version of what might happen in practice, and is given for illustrative purposes only. There are many other considerations in such decision making which cannot be mentioned here.

Other assumptions should be examined at this point, concerning the method of analysing the data. What distribution is to be used, for example, is the failure rate constant, what parameters are already known, if any. Often the assumptions are reasonable, sometimes they are made in order to make the problem tractable. Making alternative assumptions may make very little difference. The point to be made here is that formulae should not be followed blindly, but that some judgement should be used. It may be that there is really only one way to analyse the data, without going to a lot of expense and taking an extraordinary amount of time, but a result that is 80 per cent correct produced in a week is more use in a commercial situation than one that is 99 per cent correct and takes six months.

The rest of the problem is relatively easy: collect the data and perform the calculations. The intention of this text is to present these analytic techniques, and this is done, by the use of examples, in the following sections.

Examples 5.2

5.2.1 This first example is a discussion of Example 5.1. In this example, the data are already collected, a common situation. However, the initial stages should proceed as if the data were not already available, and a trial could be carried out to collect it.

First H_0 must be chosen. It could be that the null hypothesis is that the reliability is 98 per cent, or that it is *at least* 98 per cent. The conduct of the test and the analysis is different in the two cases. In order to decide, it is necessary to imagine that the data imply that the reliability may be too extreme. If it were too low, then clearly some action would have to be taken, but if it were too high, would action need to be taken then? In some situations, it could be the case that action would be taken — if the reliability is much greater than a contracted value, then possibly there could be a saving to a supplier from reducing the inspection effort, for example. In this situation, we will suppose that it is a purchaser that is conducting the test, and that he or she would not object to reliability that is too high. In that case H_0 is that the reliability is at least 98 per cent.

The significance level is a matter of choice, as was discussed earlier. As the means of making and supporting the choice of α let us suppose it is 5 per cent.

The method of analysis and the assumptions made in this case are that we are able to use the binomial distribution. The binomial assumes that the sample is chosen randomly, and that the individual trials are independent. This can be a difficult condition to meet. (Testing the next 50 items off the production line is certainly not good enough, for example. They are not chosen at random, and they may well form part of a series that is a little better, or conversely a little worse, than the average.) In practice, if a sample of items are being tested in this way, i.e. r failures out of a sample of n, then the binomial distribution, with its assumptions, is normally used. But it is as well to remember the assumptions!

The data have been collected. There are four failures in a sample of 50. The analysis consists of calculating the probability of getting a result as bad as this *or worse*. That is the probability of getting four or more failures. If we were to consider the probability of getting *exactly* the observed result, then it is not difficult to imagine situations in which every result is significant! As a thought experiment, suppose we were able to toss a million coins, and we were testing the hypothesis that the coins were fair (i.e. the probability of a head was exactly 0.5), then the probability of getting exactly 500 000 heads, is given by

$$P = C_{500\,000}^{1000\,000} (0.5)^{1000\,000}$$

$$= 8 \times 10^{-4}$$

which is significant in anybody's language. Furthermore, as this is the most probable result, whatever had happened would be significant if the probability of getting exactly the observed result were considered!

So the relevant probability is

$$p_4 + p_5 + p_6 + \ldots + p_{50} = 1 - (p_0 + p_1 + p_2 + p_3)$$

$$= 0.02$$

as stated above, using the binomial distribution with $n = 50$ and $p = 0.98$.

As

$$0.02 < 0.05,$$

i.e. our result is smaller than the significance level chosen, then the null hypotheseis is rejected, and the result is considered significant.

Note that the null hypothesis is not disproved (statisticians do not claim to have proved or disproved anything, results are just significant, or not, to varying degrees), nor is the value of the probability, 0.02 in this case, the probability that the null hypothesis is true (far from it, and this is known as the prosecutor's dilemma, which will be examined in Chapter 6). We have just calculated the prior probability of the result observed actually occurring by chance alone, if the null hypothesis is true, and we have to decide if we accept that either a rare event has taken place, or our prior assumption, the null hypothesis, is false.

5.2.2 A producer buys a large batch of components, the reliability of which he believes is 99 per cent under certain environmental test conditions. He tests 100 of them under these contitions, and three of them fail. Is this grounds for believing the batch is of poor reliability?

The null hypothesis is that the reliability is at least 99 per cent. If R is the reliability of a single item, then:

$$H_0: R > 0.99$$

The significance level is arbitrary, and in this case will be set at 1 per cent.

The data are already collected, and as the batch is large, the binomial distribution will be used.

It is necessary to calculate the probability, P, of obtaining 3 or more failures in a sample of 100 if R is 99 per cent, i.e.

$$P = P(3 \text{ or more failures})$$

$$= 1 - P(2 \text{ or fewer failures})$$

$$= 1 - (p_0 + p_1 + p_2)$$

where:

$$p_0 = R^{100} \qquad\qquad = 0.367;$$

$$p_1 = 100R^{99}F \qquad = 0.370;$$

$$p_2 = \frac{100 \times 99}{2} R^{98}F^2 = 0.185.$$

So

$$P = 1 - (p_0 + p_1 + p_2)$$

$$= 1 - 0.921$$

$$= 0.079$$

$$> 0.01$$

so the result is not significant. It is quite reasonable to assume it could have arisen by pure chance.

5.2.3 A transport company runs a fleet of vehicles. It is known that the average number of breakdowns per month is 1.5, from data collected over a number of years. One month there are 4 breakdowns. Is this significant?

The null hypothesis is that the average number of breakdowns, m, is no worse than 1.5,

$$H_0: m \leqslant 1.5$$

The significance level is 5 per cent.

Over a fleet of vehicles, the breakdowns are normally assumed to be independent, so the Poisson distribution can be used. So, if P is the probability of observing 4 or more breakdowns, then

$$P = P(4 \text{ or more breakdowns})$$

$$= 1 - P(3 \text{ or fewer})$$

$$= 1 - (p_0 + p_1 + p_2 + p_3)$$

where:

$$p_0 = e^{-1.5} \qquad\qquad = 0.223;$$

$$p_1 = 1.5e^{-1.5} \quad = 1.5p_0 = 0.335;$$

$$p_2 = \frac{1.5^2}{2} e^{-1.5} = \frac{1.5}{2} p_1 = 0.251;$$

$$p_3 = \frac{1.5^3}{3!} e^{-1.5} = \frac{1.5}{3} p_2 = 0.126;$$

So

$$P = 1 - (p_0 + p_1 + p_2 + p_3)$$

$$= 1 - 0.934$$

$$= 0.066$$

$$> 0.05.$$

The result is not significant. It could have arisen by pure chance.

5.2.4   Four identical components with constant failure rate are put on test until they fail. It is believed that the MTTF of this component design under the conditions of the test is 950 hr. The MTTF observed is 500 hr. Is this cause for concern?

The null hypothesis is that the MTTF is at least 950 hr. It is easier to work with the failure rate, λ, so the null hypothesis becomes:

$$H_0: \lambda < 1/950 = 1.05 \times 10^{-3}$$

The significance level will be 2.5 per cent in this case.

If T is the total time on test, which has been observed to be 2000 hr, then T is Erlang distributed:

$$f(T) = \frac{\lambda^4 T^3}{3!} e^{-\lambda T}$$

and the probability of obtaining a result as bad as this or worse, is given by the integral

$$\int_0^{2000} f(T) dT = 1 - \left[\frac{(\lambda T)^3}{3!} + \frac{(\lambda T)^2}{2!} + \lambda T + 1 \right] e^{-\lambda T} \Bigg|_{T = 2000}$$

by integration by parts,

$$= 0.16 > 0.025$$

and so the result is not significant.

In this case, as the distribution is Erlang, an alternative to actually doing the integrals is to use tables of the χ^2 distribution. A variable is χ^2 with v degrees of freedom if it is gamma with shape parameter $v/2$ and scale parameter $1/2$ (see Section 2.5.4).

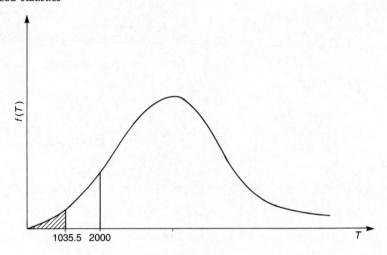

Fig. 5.1 Percentage points of the χ^2 distribution.

As T is Erlang with shape parameter 4, this corresponds to χ^2 with 8 degrees of freedom. From χ_2 tables, the lower 2.5 per cent level is 2.18 when the number of degrees of freedom is 8. That is

$$\int_0^{2.18} \frac{8x^3}{3!}\, e^{-x/2}\, dx = 0.025$$

This has $x/2$ where we want λT. Making the substitution

$$x/2 = \lambda T = T/950$$

will give us the integral we want. Changing the limits gives, when $x = 2.18$,

$$T = 950 \times x/2$$
$$= 950 \times 2.18/2$$
$$= 1035.5$$

But the value of T observed was 2000, which is far in excess of this, so the probability of getting a value of T less than 2000 must be greater than 2.5 per cent, so it is not significant.

This is illustrated in Fig. 5.1, which shows the distribution of T and the two points concerned. It can also be summed up in the following few lines. Since:

$$2000 > 1035.5$$

it follows that:

$$P(T < 2000) > P(T < 1035.5) = 0.025.$$

This is a not uncommon way to conduct tests of this nature. Rather than calculate the probability directly, the appropriate percentage points of many distributions are tabulated, and the analyst needs only to see if the value of the statistic falls on the significant side or not.

5.1.1   A sample of 50 components is from a production run that is believed to be 99 per cent reliable. On test three of them are found to be faulty. Is this evidence enough to disbelieve the assertion concerning the reliability?

5.1.2   A car hire firm with a number of depots has, on average, 2.2 breakdowns per depot per month. During one particular month one depot has 4 breakdowns. Is this evidence to believe that the maintenance at the depot is of lower quality than the average?

5.1.3   A complex piece of machinery in a production plant is believed to have an MTBF of 240 hr. During one period of 1000 hr it has 7 failures. Some of the maintainers believe that it is coming to the end of its life, and this is evidence that it is into the end of the bathtub curve and it is into its wear period. Investigate their claim using the data given.

5.4 Confidence

5.4.1 Introduction

It is not always the case that there is a hypothesized value of a parameter to use in a significance test, but it is desirable to take the statistical variability into account. This is done by the use of confidence limits. The confidence limit is the extreme value of the parameter concerned that would just not be significant at a given level were a significance test being done.

Suppose, for example, that 50 items were tested and none of them failed. Would a reasonable conclusion be that they were 100 per cent reliable? The answer is almost certainly no. In which case, what would be reasonable values for the reliability?

If the reliability were 99 per cent, then

$$P(\text{no failures}) = 0.99^{50} = 0.61 = 61\%$$

which is not at all significant.

If the reliability were 95 per cent, then

$$P(\text{no failures}) = 0.95^{50} = 0.077 = 7.7\%$$

which is not significant at the 5 per cent level.

If the reliability were 90 per cent, then

$$P(\text{no failures}) = 0.9^{50} = 0.005 = 0.5\%$$

which is significant.

If we solve the equation

$$R^{50} = 0.05,$$

we get the value

$$R = 0.942 = 94.2\%$$

and this will be the limiting value, above which the values of the reliability will not be significant, and below which they will be significant, at the 5 per cent level. It is called the 95 per cent lower confidence limit. The interval $0.942 < R < 1$ is called the lower 95 per cent confidence interval.

Note that the significance level is the 5 per cent, as it the region within which the probability is 5 per cent that is significant when doing a significance test, while it is the converse of that when finding a confidence limit, as we are 95 per cent confident that the value of R is greater than 94.2 per cent.

Note that strictly speaking, it is inaccurate to say that there is a 95 per cent probability that the true value of R lies above 0.942. This is because the true value of R is certain, and not a random variable. We cannot carry out an experiment that will verify that it is above 0.942 for 95 per cent of the time, as it is either above or it is not. However, if experimental data were analysed frequently, and 95 per cent confidence limits were calculated each time, then 95 per cent of the time, on average, the 95 per cent interval would include the true value of the parameter concerned. (Statisticians are the only professionals that demand the right to be wrong 5 per cent of the time!) Bayesian statistics (outlined in Chapter 6) allows the use of probabilities in describing certainties, and we shall return to this problem there.

5.4.2 Confidence limits on a reliability

In this section we shall examine the case outlined above, when we have a binomially distributed variable, which we shall assume is the reliability of a component or system. In practice it is very often the reliability of a one shot device.

If n such devices are tested, and r of them are observed to fail, while $n - r$ of them function, the MLE of R, the device reliability, is simply $(n - r)/n$, as would be expected. To find the confidence limit, suppose we are given a significance level α, then the $1 - \alpha$ confidence limit is required. The lower $1 - \alpha$ confidence interval is all those values of R that are not significantly small at the α level of significance, i.e. they are in the interval

$$R > R_1$$

where R_1 is the solution of

$$P(n - r \text{ or more successes in a sample of } n) = \alpha.$$

Example 5.3 Suppose 99 successes are observed in a sample of 100. Then the 95 per cent upper confidence limit is the solution of the equation

$$R^{100} + 100R^{99}(1 - R) = 0.05$$

which is (to three places of decimals)

$$R = 95.3\%$$

Notice that this is the lower confidence limit. The upper confidence limit is the solution to to equation

$$P(n - r \text{ or fewer successes in a sample of } n) = \alpha$$

or, turning the probability round,

$$P(\text{more than } n - r \text{ successes}) = 1 - \alpha$$

which in this example is the solution of

$$R^{100} = 0.95$$

giving

$$R = 0.999.$$

The interval between 0.953 and 0.999 is a two sided interval, and this is the 90 per cent two sided confidence interval. Note that it is only 90 per cent as there is a 5 per cent significant tail above R and a 5 per cent significant tail below R.

5.4.3 Confidence limits on an MTTF

Suppose that a number of items are put on test until they fail. Then, assuming a constant failure rate, the best estimate of the MTTF is the total time on test, T, divided by the number of failures, n, T/n. In order to calculate the confidence limits, it is necessary to examine the distribution of T, which is Erlang with shape parameter n and scale parameter λ. Then

$$f(T) = \frac{\lambda^n T^{n-1}}{(n-1)!} e^{-\lambda T}.$$

The $1 - \alpha$ upper confidence on limit on λ, λ_u, is the solution to the equation

$$\int_0^T f(t)\, dt = \alpha$$

for λ.

In practice, although the integral can be done analytically, as was illustrated when calculating the significance in Section 5.3, this would necessitate having to solve a difficult equation, and it is easier to use the χ^2 distribution, as was done as an alternative in 5.3. It is best illustrated by means of an example.

Examples 5.4

5.4.1   It is easier to take data that has already been examined, and so the example will use the data from the example in Section 5.3.1, namely that 4 components were tested to failure, and the observed MTTF was 500 hr. This means that the total time on test was 2000 hr.

If we were to solve the integral analytically, this would give the equation

$$\int_0^{2000} f(T)\,\mathrm{d}T = -\left[\frac{(\lambda T)^3}{3!} + \frac{(\lambda T)^2}{2!} + \lambda T + 1\right] e^{-\lambda T}\bigg|_{T=2000}$$

$$= 0.025$$

to be solved for λ.

Far easier is to use the χ^2 distribution. Note from the tables that the 2.5 per cent point of the χ^2 distribution with 8 degrees of freedom is 17.53. In order to change from the χ^2 distribution, it is necessary to make the change of variable

$$x/2 = \lambda T.$$

This time we know the value of T, but not the value of λ. So, solving for λ

$$\lambda = \frac{x}{2T} = \frac{17.53}{2 \times 2000}$$

$$= 43.8 \times 10^{-4}$$

which corresponds to an MTTF of 228 hr. This is the 97.5 per cent lower confidence limit on the MTTF.

To find the upper confidence limit on the MTTF, or the lower confidence limit on λ, it is necessary to solve the equation

$$\int_{2000}^{\infty} f(T)\,\mathrm{d}T = \alpha$$

or alternatively, the equivalent equation,

$$\int_0^{2000} f(T)\,\mathrm{d}T = 1 - \alpha$$

and this can be done by the use of χ^2 tables. To find the 97.5 per cent upper confidence limit on the MTTF, note that from χ^2 tables, the upper 97.5 percentage point for χ^2 with 8 degrees of freedom is 2.18. Then applying the transformation

$$x/2 = \lambda T$$

as before, this gives

$$\lambda = \frac{2.18}{2 \times 2000}$$

$$= 5.45 \times 10^{-4}$$

which corresponds to an MTTF of 1835 hr.

In summary, the point estimate of the MTTF is 500 hr, and the upper 97.5 per cent confidence limit on the MTTF is 1835 hr, and the lower 97.5 per cent confidence limit is 228 hr. The 95 per cent two sided interval is from 228 hr to 1834 hr.

5.4.2 Twenty components were tested until failure, and the observed MTTF was 500 hr. Calculate the 95 per cent two sided confidence interval on the MTTF.

The two values of χ^2 with 40 degrees of freedom, at the 2.5 percentage and 97.5 percentage points are from tables, 24.43 and 59.35, and the total time on test is 10 000 hr. To find the limits on the MTTF, it is necessary to solve the equation

$$x/2 = \lambda T$$

for λ, substituting the two values of x given above, remembering that $\lambda = 1/\text{MTTF}$. This gives the values 819 hr and 337 hr, so the 95 per cent confidence interval is from 337 hr to 819 hr. Notice that because there is so much more data than in the previous example, although the point estimate of the MTTF is the same, the confidence interval is so much shorter, as would be expected.

If instead we had wanted the 99 per cent confidence interval, then the 99.5 percentage and 00.5 percentage points of the χ^2 distribution are 20.71 and 66.77. This gives 966 hr and 300 hr, so the 99 per cent interval is from 300 hr to 966 hr. In order to have more confidence, the interval must be wider.

Exercises 5.2

5.2.1 In a trial, 40 components were tested and one failed. Calculate the 95 per cent two-sided confidence interval.

5.2.2 A sample of 10 components was tested until they had all failed. The total time on test was 2300 hr. Calculate the 99 per cent two-sided confidence interval on the MTTF.

5.2.3 In a period of one year, a complex system failed 9 times. What is the 90 per cent lower confidence limit on the MTBF?

5.5 Goodness of fit tests

5.5.1 Introduction

It is frequently the case that the assumption that the data come from a given distribution must be examined, i.e. how well the data fits the distribution. As was seen at the end of Section 4.4.3, the use of the wrong distribution can lead to the wrong conclusion. We shall examine two tests; one based on the χ^2 distribution, the other the Kolmogorov-Smirnov test.

5.5.2 The chi-squared test

This test is applicable when many data are available. If the data are sparse, it cannot be used. It is best illustrated by an example.

Examples 5.5

5.5.1 Suppose the number of repairs at a number of depots during a long time period is recorded, and the frequency of repairs in each month at each depots summarized in Table 5.1. Do the repairs occur randomly? Putting it into the language of statistics, could the data have come from a Poisson distribution?

Table 5.1 Frequencies of the numbers of repairs in different months

Number of repairs	0	1	2	3	4	5	6 or more
Frequency	16	17	13	10	3	1	0

The mean of the data is 1.5. We shall compare the data with a Poisson distribution with mean 1.5, and perform a test to see if there is a significant difference between the data and the expected values from the Poisson.

If the data are Poisson with mean, $m = 1.5$ then the probability of observing r repairs in a month is

$$p_r = e^{-m} \frac{m^r}{r!}$$

and the expected number of months in which there are r repairs in a total of N months is given by

$$e_r = N p_r = \frac{m}{r} e_{r-1}.$$

The values of e_r are shown in Table 5.2, along with the observed values shown in Table 5.1. The equation for χ^2 is given by

Table 5.2 Observed and expected frequencies of repairs

Number of repairs	0	1	2	3	4	5	6 or more
Observed frequencies	16	17	13	10	3	1	0
Expected frequencies	13.4	20.1	15.1	7.5	2.8	0.9	0.4
Combined expected frequencies	13.4	20.1	15.1			11.4	
Combined observed frequencies	16	17	13			14	

$$\chi^2 = \sum \frac{(o - e)^2}{e}. \tag{5.2}$$

There is just one further proviso before the value of χ^2 can be calculated: this is actually an approximation only, and if the value of e, the expected value of the frequency, falls below 4, then the errors in the approximation become too great. For that reason, some of the values of e, and also the corresponding values of o, are combined until they are greater than 4. This is also shown on Table 5.2.

Applying the formula gives

$$\chi^2 = 1.87.$$

The number of degrees of freedom is obtained by first of all counting up the number of observations left after we have combined some of them to ensure that the values of e are all over four, giving four in this case, and then subtracting one for each parameter that has been obtained from the data. In this case there are two parameters, namely $N = 60$, the total number of data, and $m = 1.5$, the mean of the Poisson. This leaves two degrees of freedom.

Looking up the χ^2 tables, the value of χ^2 corresponding to 10 per cent significance, when the number of degrees of freedom is two, is 4.61. This is far in excess of the value obtained from the data, and so the result is not significant.

It may be asked how we know which side of the value obtained from the table is the significant one. In the best of all possible worlds, where the fit was perfect, χ^2 would be zero (or at least very small, as the observed values must be integers, and the expected values will almost certainly not be), and the larger the value of χ^2 observed, the more reason to believe that the data does not come from a Poisson distribution. The values in the tables give the limiting values for a given degree of worry (the significance level).

5.5.2 Test the life data of Table 2.3 to see if the component in question could have a constant failure rate.

The mean of the data is 8.522 hr. This corresponds to a constant failure rate of 0.117 failures/hr. The null hypothesis is:

H_0: the data is from an exponential distribution with mean 8.522 hr.

Table 5.3 Grouped life data showing expected and observed values, and combined cell values to ensure expected values all exceed four

Interval	Observed frequency	Expected frequency	Grouped Observed	Grouped Expected
1	2	11.1	2	11.1
2	3	8.8	3	8.8
3	5	7.8	5	7.8
4	3	6.9	3	6.9
5	9	6.2	9	6.2
6	11	5.5	11	5.5
7	8	4.9	8	4.9
8	13	4.3	13	4.3
9	9	3.9	16	7.3
10	7	3.4		
11	5	3.0	11	5.7
12	6	2.7		
13	4	2.4	8	4.5
14	4	2.1		
15	3	1.9	5	5.1
16	0	1.7		
17	2	1.5		
18	1	1.3		
19	1	1.2	3	4.5
20	1	1.1		
21	0	0.9		
22	0			
23	2		3	17.4
24	0			
25	1			

The probability of a data point falling in the interval (t_1, t_2) is given by the integral

$$P(t_1, t_2) = \int_{t_1}^{t_2} \lambda e^{-\lambda t}\, dt$$

$$= e^{-n\lambda}(1 - e^{-\lambda})$$

when t_1 and t_2 are n and $n + 1$ respectively, in which case $P(n, n + 1)$ is the probability, P_{n+1} of the data point falling in the $(n + 1)$th interval. If we have N data points (100 in this case), then the expected number of data points, e_n, in the nth interval is NP_n. Table 5.3 shows the values of e_n, and the values after combining enough of the data points to ensure that all the expected values are greater than four. This gives

$$\chi^2 = 69.66.$$

There are 14 data points after combining the data, and the total number of data points and the mean are both obtained from the data, leaving 12 degrees of freedom. Looking up χ^2 with 12 degrees of freedom, the largest value in the table is 32.91, at the 0.1 per cent significance level. This is highly significant, and it is very unlikely that the data comes from an exponential distribution.

5.5.3 The Kolmogorov–Smirnov test

This test for goodness of fit is an alternative to the χ^2, with advantages and disadvantages. It does not depend on grouping the data, which can be arbitrary, and is applicable to small samples which would not be tractable using the χ^2 test. On the other hand, for large samples it involves more work, unless a computer is available.

The test looks at the difference between the cumulative function, $F(t)$, of the hypothesized distribution and the observed cumulative function, which we will call $F_0(t)$. The value used is the maximum value observed. The distribution of this value is independent of the parent distribution, but does depend on the sample size. If the fit was exact, then this value would be zero, so large values give cause for concern. Tables of the critical values of the distribution are available in many textbooks and books of tables.

Example 5.6

5.6.1 Ten items are put on test, and their lifetimes are recorded. It is believed that the item has a constant failure rate. The null hypothesis is:

H_0: The data is from an exponential distribution with MTTF 100 hr.

The data and the analysis are shown in Table 5.4.

Table 5.4 Kolmogorov–Smirnov analysis of hypothesized exponential data

Failure number	Lifetime	Median rank	Theoretical F value	Absolute difference
1	3	0.067	0.030	0.037
2	20	0.163	0.181	0.018
3	40	0.260	0.330	0.070
4	52	0.356	0.405	0.049
5	53	0.452	0.411	0.041
6	54	0.548	0.417	0.131
7	85	0.644	0.573	0.071
8	318	0.740	0.958	0.218
9	429	0.837	0.986	0.150
10	553	0.933	0.996	0.063

The maximum value of the absolute difference is the 8th, and is 0.218. This is less than the 10 per cent significance value from the table, which is 0.369, so the result is not significant at the 10 per cent level.

The values of $F_0(t)$ used here are the median rank values, presented in eqn 4.6 but other authors use mean rank values, while others again use i/n for the value of $F_0(t_i)$. See DeGroot (1975), Grosh (1989) and O'Connor (1991) for more discussion on this point.

Exercises
5.3

5.3.1 The data in Table 5.5 are believed to come from an exponential distribution. Use the χ^2 test to examine this hypothesis.

Table 5.5 Hypothesized exponential data

Time (hr)	0–1	1–2	2–3	3–4	4–5	5–6	>6
Frequency	12	8	7	4	3	1	0

5.3.2 In a trial, 24 electronic components were run until they failed. Their lifetimes are shown in Table 5.6. Test the hypothesis that they have a constant failure rate.

Table 5.6 Hypothesized exponential data

111	89	233	24	6	49
28	169	346	149	191	207
189	405	52	42	44	243
142	135	79	121	575	39

5.3.3 In another trial, 30 bearings were run until failure. Their lifetimes, shown in thousands of revolutions, are shown in Table 5.7. It is believed that the failures were due to wear, with no minimum life. Test the hypothesis that the data is from a Weibull distribution.

Table 5.7 Hypothesized Weibull data

73	122	123	91	39	48
42	56	184	28	111	99
64	50	105	68	68	137
88	47	20	99	32	141
115	108	62	81	175	61

5.3.4 The data of Table 5.8 is believed to come from an exponential distribution. Use the Kolmogorov–Smirnov test to examine this hypothesis.

Table 5.8 Hypothesized exponential data

238	491	45	672	926	101

5.3.5 Six bearings were tested to failure. It is believed that their lifetimes are Weibull distributed. Use the Kolmogorov–Smirnov test to examine this hypothesis. The data are shown in Table 5.9.

Table 5.9 Hypothesized Weibull data

6.4	28	24	16	253	397

5.6 Contingency tables

Suppose a manufacturer buys components from three different suppliers. In order to test the quality, he tests samples of the components from each of the suppliers under stringent conditions, and each component either fails or survives its test. The data are shown in Table 5.10. Such a table is called a contingency table.

It is required to test if the apparent difference in the reliabilities of the components from the different suppliers comes from a real difference, or could be put down to pure chance, i.e. it tests if the reliability is independent of the supplier. The test described here is the χ^2 test, and proceeds as follows.

Table 5.10 is the table of observed results. The table of expected results assumes that the reliabilities of the three suppliers is the same, so that the null hypothesis is:

H_0: There is no diference between the suppliers.

Table 5.10 Number of functioning and failed components from different suppliers (observed)

	Suppliers			
	A	B	C	Totals
Functioned	90	72	58	220
Failed	10	8	12	30
Totals	100	80	70	250

Table 5.11 Expected number of functioning and failed components from different suppliers

	Suppliers			
	A	B	C	Totals
Functioned	88.0	70.4	61.6	220.0
Failed	12.0	9.6	8.4	30.0
Totals	100.0	80.0	70.0	250.0

If this is the case, then the average overall reliability, 220/250, is the best estimate of the common value of the reliability, and if this were the value, then the expected values of the numbers of successes and failures would be the values given in Table 5.10. The numbers in the columns are obtained by dividing the column totals, i.e. the number tested from each manufacturer, in the ratio 88:12, which is the ratio of successes to failures overall (220/250 = 0.88).

The mathematical formula is as follows. If t_i is the total of the ith row, and s_j the total of the jth column, then the expected value e_{ij} in the box which is the intersection of the ith row and jth column is given by

$$e_{ij} = \frac{t_i s_j}{T} \qquad (5.3)$$

where T is the grand total.

The value of χ^2 is given by the formula in eqn 5.2, where the o values are from Table 5.10 and the e values from Table 5.11. This gives

$$\chi^2 = \frac{(90 - 88)^2}{88} + \frac{(72 - 70.4)^2}{70.4} + \frac{(58 - 61.6)^2}{61.6} + \frac{(10 - 12)^2}{12} +$$

$$\frac{(8 - 9.6)^2}{9.6} + \frac{(8.4 - 12)^2}{8.4}$$

$$= 3.57.$$

The number of degrees of freedom, v, is calculated from the following formula:

$$v = (n - 1)(m - 1)$$

where n is the number of columns and m the number of rows. So v is 2 in this case. From χ^2 tables, the critical value of χ^2 at the 10 per cent significance level when $v = 2$ is 4.61. The observed value of 3.57 is less than that, so our result is not significant, i.e. there is no evidence to suggest that the suppliers are supplying at different reliabilities.

Examples 5.7

5.7.1 A user believes that light bulbs last for about 2000 hr. He wishes to test a variety of new designs against the design he has been using in the past. He then buys a number of the old and new designs, and records their lifetimes over several months. In order to simplify the data collection, the lifetime was deemed to be short, less than 1800 hr, average, 1800 to 2200 hr, or long, greater than 2200 hr. The data are summarized in Table 5.12. Does this show any evidence that any of the new designs last longer than the old one?

The null hypothesis is:

H_0: there is no difference between the light bulbs.

Table 5.12 Light bulb data (observed)

	Design				
	Old	New			
		1	2	3	Totals
Short	29	40	38	19	126
Average	529	611	423	558	2121
Long	17	32	28	51	128
Totals	575	683	489	628	2375

Table 5.13 Light bulb data (expected)

	Design				
	Old	New			
		1	2	3	Totals
Short	30.5	36.2	25.9	33.3	126.0
Average	513.5	610.0	436.7	560.8	2121.0
Long	31.0	36.8	26.4	33.8	128.0
Totals	575.0	683.0	489.0	628.0	2375.0

Table 5.12 shows the observed results. The expected results are shown in Table 5.13. They are obtained using eqn 5.3. The value of χ^2 obtained using eqn 5.2 is

$$\chi^2 = 28.86.$$

The value of χ^2 from the table, with 6 degrees of freedom is 22.46 at the 0.1 per cent level of significance. So the result is very significant, and there is evidence to suggest the designs are different.

5.7.2 It is believed that the use of new technology has improved the ability of cars to start in the winter, to the extent that if properly maintained, there is no difference between winter and summer. A car hire company decided to test this claim, and recorded the number of times its cars failed to start first thing in the morning during the year. The results are shown in Table 5.14.

The null hypothesis is:

H_0: there is no difference between summer and winter.

Table 5.15 shows the expected results, calculated using eqn 5.2. The value of χ^2 is

$$\chi^2 = 4.50.$$

Examining the tables of χ^2 percentage points, with $v = 1$, the critical value at the 5 per cent significance level is 3.84, and as the observed value

Table 5.14 Car starting data (observed)

	Technology		
	Old	New	Totals
Started	1589	1572	3161
Did not start	238	189	427
Totals	1827	1761	3588

Table 5.15 Car starting data (expected)

	Technology		
	Old	New	Totals
Started	1609.6	1551.4	3161
Did not start	217.4	209.6	427
Totals	1827	1761	3588

is greater than this, the result is significant at the 5 per cent level. However, the observed value is less than 5.02, the critical value at the 2.5 per cent significance level, so the result is not significant at the 2.5 per cent level.

Exercises 5.4

5.4.1 A producer wished to compare two suppliers products. Over a long period he examined the quality of the the components he received from them. The data he collected are shown in Table 5.16. Does this data show evidence that there is a difference between the suppliers?

Table 5.16 Comparison of supplier reliability

	Supplier	
	A	B
Number of components functioning	382	571
Number of components failing	17	21

5.4.2 A retailer with four outlets was interested in comparing the quality of the goods sold at the different outlets. He recorded the total sales and the number of warranty claims from each outlet, of one item of stock during one week. The data are shown in Table 5.17. Does this data show evidence that the quality of the items sold varies from outlet to outlet?

5.4.3 The management of a factory believe that a new manufacturing technique can improve their quality. They run a trial for a week, during which every item produced at the factory is examined, and the two

Table 5.17 Warranty claims comparison of different outlets

	Outlet			
	A	B	C	D
Total sales	520	375	438	102
Warranty claims	48	41	42	6

Table 5.18 Comparison of manufacturing techniques

	Technique	
	Old	New
Acceptable	1082	1872
Unacceptable	48	63

techniques compared. The data are shown in Table 5.18. Does it support the hypothesis that there is a diffence between the old and new techniques?

5.7 Linear regression

5.7.1 Introduction and basic formulae

Linear regression is sometimes called least squares fit, and is concerned with finding lines of best fit, as was done during Weibull analysis and Duane plotting. This section deals with this analysis.

It is believed that increasing one of the dimensions of a component of a water tap will increase the life of the washer. In order to test this theory, ten taps with ten different dimensions are made, and tested under standard conditions in which the taps were continually turned on and off. The lives of the washers were recorded. The data pairs, dimension and washer life, are shown in Table 5.19.

The easiest way to illustrate if there might be a connection between theses parameters is to draw a scatter diagram, as shown in Fig. 5.2. Each point on the chart represents one trial, with x-coordinate the critical dimension, and y-coordinate the corresponding life. By examining the diagram, it appears that increasing the dimension may increase the life. There appears to be a trend, despite the fact that there is not a uniform increase in the data.

Suppose there is a line, with equation

$$y = ax + b$$

and it is required to find the 'best' values of a and b, where best is to be defined.

Table 5.19 Washer life and tap dimension

Tap number	Dimension (mm)	Life (days)
1	10.0	122
2	10.1	134
3	10.2	105
4	10.3	141
5	10.4	168
6	10.5	159
7	10.6	160
8	10.7	178
9	10.8	153
10	10.9	172

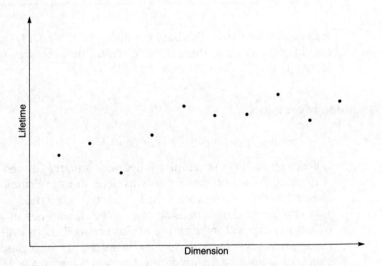

Fig. 5.2 Scatter plot of lifetimes as a function of dimension.

The commonest way is the least squares approach, which works as follows. Suppose the vertical distance between each point and the line is considered, that is the distance, ε_i for the ith point, given by

$$\varepsilon_i = y_i - (ax_i + b)$$

called the error. The ε_i are shown in Fig. 5.3. Then the sum of squares of errors is minimized, i.e. the expression

$$V = \sum \varepsilon_i^2$$

$$= \sum (y_i - (ax_i + b))^2$$

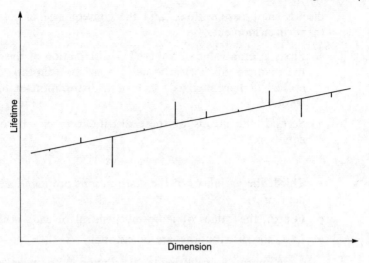

Fig. 5.3 Scatter plot showing regression line and errors.

is minimized. The minimization is an exercise in differential calculus, the solution of which is

$$a = \frac{N \sum x_i y_i - \sum x_i \sum y_i}{N \sum x_i^2 - \left(\sum x_i\right)^2} \qquad (5.4)$$

where N is the number of data points, and

$$b = \bar{y} - a\bar{x} \qquad (5.5)$$

For the data in the example, the values of a and b are given by

$$a = 60.4$$

$$b = -481.6$$

and the line is shown in Fig. 5.3.

5.7.2 The statistics

Equations 5.4 and 5.5 have little to do with statistics. There is clearly some variability in this situation. If the experiment were to be repeated with ten more taps, of the same dimensions as the originals, then the lifetimes would clearly be different, giving different values for a and b, so that a and b are distributed. It is dealing with that variability in a and b that is the purpose of this section.

The statistical analysis makes the following assumptions. We shall discuss these assumptions, and the consequence of relaxing some of them, in a moment.

- First, at each value of x there is a distribution of the values of y. In the example this is clearly the case, as for each tap design, with the critical dimension fixed, there is a distribution of lifetimes of the washer.
- Second, the means of these distributions, μ_y, lie on a line with equation

$$\mu_y = \alpha x + \beta. \tag{5.6}$$

- Third, the variances of the distributions are identical (i.e. independent of x).
- Fourth, the values of y are independent of each other.
- Last, the distributions are all normal.

In summary, y is normally distributed with mean μ_y defined by eqn 5.6 and variance σ^2, where σ^2 is independent of x. From this can be deduced (see Ott(1984), for example) the following facts about a, b, and σ.

The distribution of a is normal, with mean α and variance σ_a^2 given by

$$\sigma_a^2 = \frac{\sigma^2}{\sum x_i^2 - \left(\sum x_i\right)^2 / N}$$

and the distribution of b is normal, with mean β and variance σ_b^2 given by

$$\sigma_b^2 = \frac{\sigma^2 \sum x_i^2}{N \sum x_i^2 - \left(\sum x_i\right)^2}.$$

If the Sum of Squares of Errors, the SSE, is given by

$$\text{SSE} = \sum \varepsilon_i^2 = \sum (y_i - (ax_i + b))^2$$
$$= (N - 1)(s_y^2 - a^2 s_x^2)$$

then an unbiased estimate of σ^2, s_ε^2, is given by

$$s_\varepsilon^2 = \frac{\text{SSE}}{N - 2}.$$

The $(1 - \gamma)$ confidence intervals on α and β are given by the equations

$$a \pm \frac{t_{\gamma/2} s_\varepsilon}{\sqrt{\sum x_i^2 - \left(\sum x_i\right)^2 / N}}$$

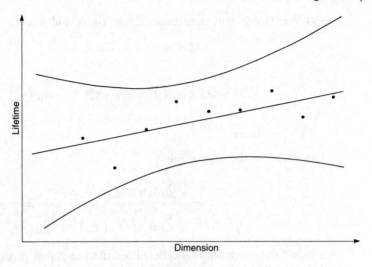

Fig. 5.4 Scatter plot showing regression line and its confidence limits.

$$b \pm t_{y/2}s_{\varepsilon} \sqrt{\frac{\sum x_i^2}{N \sum x_i^2 - \left(\sum x_i\right)^2}}$$

where $t_{y/2}$ is the critical value of the t distribution with $N - 2$ degrees of freedom.

It is also possible to put a confidence interval on the line itself. The equation for the limits is

$$y = ax + b \pm t_{y/2}s_{\varepsilon} \sqrt{\frac{1}{N} + \frac{\sum (x_i - x)^2}{\sum x_i^2 - \left(\sum x_i^2\right)^2 / N}}.$$

The use of these is illustrated using the data of the example. The confidence limits on the line are shown in Fig. 5.4.

We already had

$$a = 60.4$$

$$b = -481.6$$

then the SSE is given by

$$SSE = 1915.5$$

and

$$s_{\varepsilon} = 15.5$$

and the 95 per cent confidence limits on α and β are

$$60.4 \pm 32.4 = (28.0, 92.8)$$

and

$$-481.6 \pm 410.7 = (-829.3, -70.9).$$

5.7.3 Correlation

The value of the expression

$$r = \frac{\sum x_i y_i - \sum x_i \sum y_i / N}{\sqrt{\left(\sum x_i^2 - \left(\sum x_i\right)^2 / N\right)\left(\sum y_i^2 - \left(\sum y_i\right)^2 / N\right)}}$$

is called the correlation coefficient of the sample. It gives a measure of the strength of the linear relationship betwen the two variables. If r is plus or minus one, then the points all lie on a straight line, while if r is zero, there is no relationship between them. If r is positive, then y tends to increase with x, while if r is negative, y tends to decrease as x increases.

If we are using the regression line to try to predict the value of y from a knowledge of x, then r is a measure of the amount of information gained about y as a result of having knowledge about x. In statistical language, there is variability in y, which can be traced to the variability in x, and the variability in the distribution of y given x (measured by σ^2 in the previous section). The value r^2 gives the proportions in which the total variability in y is divided between these two.

For the data we used in the example

$$r = 0.78.$$

Exercise 5.5

5.5.1 The data in Table 5.20 comes from a trial in which 10 gearboxes were tested, at different power outputs, until failure. It shows the life-times and the power outputs of each gearbox. Draw a scatter diagram of the data showing how the lifetimes vary as a function of the power output. Find the regression line, and put it on your drawing. Find the SSE, and hence put the 95 per cent confidence limits on α and β. Find the correlation coefficient.

5.7.4 The assumptions

The assumptions mentioned above will be discussed here, particularly with respect to the Weibull distribution, and the estimation of the parameters. In Weibull analysis, the time, t, is the dependent parameter, despite being on the horizontal axis rather than the vertical. In plainer terms, and the author has heard it expressed this way, the results are

Table 5.20 Power output and life of gearboxes

Gear box number	Power output	Life ($\times 10^{-4}$)
1	100	104
2	200	92
3	300	65
4	400	72
5	500	76
6	600	69
7	700	54
8	800	39
9	900	37
10	1000	25

better if the least squares is done horizontally rather then vertically as is normal.

The assumption regarding the means lying on a straight line holds for the Weibull, although when Weibull analysis was presented (Section 4.4) the median values were used. The means were given, but these are approximations and White (1967) gives the true mean values.

The variances of the distributions are certainly not identical. When this is the case, this can be taken care of if there is sufficient knowledge concerning how the variances change with the values of x. The sums, and particularly the SSE, must now take the sums weighted inversely by the variances (so that more notice is taken of those points with small variances). In the case of the Weibull distribution, the variances can also be calculated, and White (1967) also gives the values of the variances in his paper.

The normality assumption is also not true for the Weibull distribution, but the deviation from the normal is not too severe, and this will not affect the results too much. The normality assumption will also matter less and less as the number of data points increases.

The independence assumption certainly doesn't hold as the times are ordered, i.e. it is known in advance that

$$t_i < t_j$$

whenever

$$i < j.$$

This too, can be taken care of, and White (1967) discusses this in his paper.

The effect of ignoring these difficulties is not catastrophic. In crude terms, the more information we have about a situation before the trial takes place and the data are collected, the more confidence there will be in the final result. In statistical terms, this is reflected in the confidence interval, which will be narrower when these effects are taken into account. White (1967) also discusses this point, and gives an example.

Some advanced techniques

6

6.1 Introduction

This book deals mainly with the commoner analytic techniques; there are naturally many more, some of which are dealt with in this chapter. Two, Markov analysis and Bayesian statistics, are dealt with in more depth in subsequent sections. In this section other techniques are discussed briefly.

6.1.1 Competing risks

When analysing life data, it is sometimes the case that the failure modes of the failures observed can be identified, in which case it can be useful to know the statistical properties of the failure modes. For example, a bearing might fail due to wear or fatigue, with both failure modes being approximately equally represented in a given data set. What is required is knowledge of the wear characteristics if there were no fatigue, and also the fatigue characteristics if there were no wear.

In this example, the two risks are wear and fatigue. If we wished to simulate this situation, the simplest way would be to generate two random times, one a wear time, and one a fatigue time, and the simulated time would be the minimum of these. In order to be able to perform the simulation, we must assume that the pdfs of the appropriate lifetimes are each known as if the other were not present, and it is estimating these pdfs that is the problem addressed by the theory of competing risks.

The problem arises because each lifetime comes from one of the pdfs, but gives a censored reading for the other. This is a problem that was partly addressed in Section 4.4 on Weibull analysis, Subsection 4.4.4 on censored data, but it is not always convenient to use the Weibull distribution to analyse such data. See Crowder *et al.* (1991), and Mann *et al.* (1974), for more information on this topic.

6.1.2 Proportional hazards

In a number of situations data come from different, although similar, sources, and it is desired to use the information in the different trials as inputs into the others in some way. One way is to assume that the failure rates are always in a constant proportion, in which case the proportion has to be one of the parameters to be estimated.

A common example is accelerated testing, when items are tested in a variety of environments, usually harsher than those they would meet in service, in order to induce failures and obtain some information concerning their likely life characteristics. If it is assumed that the only effect of the harsher environment is to multiply the failure rate (which is not assumed constant) by a fixed amount, so that the failure rates are in a fixed proportion, which in the simplest model is invariant with time, then information from the harshest environments can be used in the analysis of the data from the normal environment (which is almost certainly very sparse).

The different environments are called covariates, and a number of covariates can be studied at once, including their combined effects. An example could be to study the effect of different manufacturing techniques, different temperatures, and different modes of vibration during use. The technique was first developed by Kalpfleisch and Prentice (1980) in order to study human mortality statistics obtained from cancer patients, in which case the covariates would be such things as different treatments, different ages and sexes of the patients, different parts of the country, etc. See also Crowder *et al.* (1991) for more information.

6.1.3 Load strength interference modelling

This is an engineering analysis that attempts to deal with failures at source, by assuming that the loads experienced by a component are not fixed, but are distributed, as are the strengths of the components. This leads to two distributions that have to be considered, one of loads and one of strengths, and the conclusion that unreliability comes from the region where they overlap.

Carter (1986), discusses this, in the context of design of mechanical components, and compares the technique with the minimum strength/maximum load technique used classically. His results are interesting, in that he confirms some commonly used practices, but is scornful of the factor of safety, or factor of ignorance, as he has called it.

Further results are given by Leitch (1990), when he discusses the effects of interference on the product rule for reliabilities for components in series, given in Section 1.8.2 (eqn 1.2). The result is that the system reliability lies between the product of the component reliabilities and the reliability of the least reliable component. Exactly where depends on the parameters of the distributions concerned, and both Carter's and Leitch's results depend on doing some fairly gruesome multiple integrals, but they are interesting from a theoretical point of view, and the interested reader should read the original papers, which will give him or her some idea of the statistical structure behind some of the phenomenon that may be observed.

6.2 Markov modelling

6.2.1 Introduction

Markov processes are useful in reliability for modelling systems in which there are several, degraded states, and for taking into account some of the difficulties that are overlooked with other techniques, such as added stresses when redundant elements fail, dormant failure, varying numbers of repair teams and repair policies, etc., which cannot be modelled using RBDs. Naturally, as the models deal with more assumptions, the mathematics is more complex, but computers can now deal with many of the problems that arise from having to solve the equations that arise.

A situation is said to be Markovian when future states depend only on the present state and not on the past (how long the system has been in the present state, or the previous state). One good example of a Markov process is the weather. The weather tomorrow is most likely to be identical to that of today, independent of the weather yesterday, or how long the weather has been like it is.

Markov processes can be discrete or continuous, and we shall look at both cases, starting with the discrete situation.

6.2.2 The discrete case

Suppose that a complex system can be in one of three states, called fully functioning, degraded, and failed. Call them S_1, S_2, and S_3. Then the Table 6.1 shows the probability of the system being in state i tomorrow given that it is in state j today, $i, j = 1, 2, 3$. An example of such a system could be a computer system, consisting of a single, central computer with a number of terminals. The specification for the system gives priority for some users over others. The system is said to be fully functioning if everybody, that is, high and low priority users, have access, and is

Table 6.1 Table of transition probabilities

		Today		
		S_1	S_2	S_3
	S_1	0.95	0.30	0.20
Tomorrow	S_2	0.04	0.65	0.60
	S_3	0.01	0.05	0.20

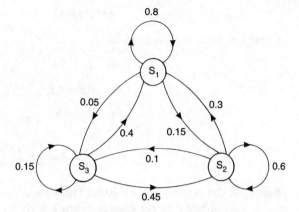

Fig. 6.1 State space diagram of computing system.

degraded if there is only enough functionality for the high priority users to be able to use it. It is failed if it does not even meet this requirement.

The matrix of probabilities is called a probability transition matrix, or Markov matrix. It has the property that its entries are non-negative, and that the columns sum to one. The matrix summarises all the information about the system. (Note: there is no universal convention about which way up the transition matrix should go. Some authors follow the convention given here, while some have the 'Today' states down the left-hand side and the 'Tomorrow' states along the top. In the latter case, the roles of the rows and columns are interchanged, and the rows must sum to one.)

An alternative way of representing the system is the diagram shown in Fig. 6.1, called a state space diagram or a Markov diagram. It has on it all the information that is in the transition matrix, and is equivalent to it in the sense that each can be derived from the other.

Suppose it is required to find the long term availability of the system. Then let P_i be the probability that the system is in state i, $i = 1, 2,$ and 3. Then the P_i are the solutions of the linear equations:

$$P_1 = 0.95P_1 + 0.30P_2 + 0.20P_3;$$

$$P_2 = 0.04P_1 + 0.65P_2 + 0.60P_3;$$

$$P_3 = 0.01P_1 + 0.05P_2 + 0.20P_3;$$

$$1 = P_1 + P_2 + P_3.$$

The derivation of these equations is illustrated for a system with three states. The technique is easily extended to the general case with n states. Suppose we have a system with three states, and probability transition matrix M, where

$$M = (a_{ij})$$

so that the transition matrix is

$$
\begin{array}{ccc}
a_{11} & a_{12} & a_{13} \\
a_{21} & a_{22} & a_{23} \\
a_{31} & a_{32} & a_{33}
\end{array}
$$

and

$$\sum_i a_{ij} = 1 \text{ for each } j.$$

Then a_{ij} is the probability of going from state j to state i, when $i \neq j$, and the probability of no change when $i = j$.

If the system is in state j at time $n + 1$, then it must been in one of the three states 1, 2, or 3 at time n and suffered a change. This is illustrated in the Fig. 6.2. If A_i is the long term probability of being in state i, the above becomes

$$A_j = a_{j1}A_1 + a_{j2}A_2 + a_{j3}A_3.$$

For example, if $j = 1$, the equation says that the probability of being in state 1 at time $n + 1$ is the probability of being in state 1 at time n and not changing, plus the probability of being in state 2 at time n and suffering a change, plus the probability of being in state 3 at time n and suffering a change. This is simply an application of the product and summation rules that were discussed in Chapter 2.

Notice how the first three equations are obtained from the transition matrix. In matrix notation, if P is the vector of probabilities, then the first three equations can be written

$$P = MP$$

or

$$(M - I)P = 0$$

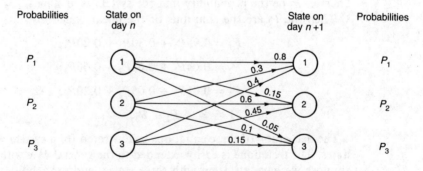

Fig. 6.2 Derivation of equations for long term availability.

where I is the identity matrix. Being able to solve this equation depends on one of the eigenvalues of M being one, and it can be shown that a transition matrix will always have unity as one of its eigenvalues. (For a definition and description of eigenvalues and eigenvectors, and an account of their properties, see any good undergraduate text on linear algebra.)

The fourth equation must be added to ensure that the probabilities sum to one. This leads to the apparent anomaly of having four equations but only three unknowns, but on close examination, it will be seen that one of the first three equations is redundant, and can be obtained from the other two.

The solution to these equations in this case is

$$P_1 = \frac{59}{96} = 0.62$$

$$P_2 = \frac{5}{16} = 0.31$$

$$P_3 = \frac{7}{96} = 0.07.$$

That is, the system will be in the fully functioning state for 62 per cent of the time in the long term, and if the degraded state is acceptable, the availability will be 93 per cent. If these figures are not acceptable, then the entries in the transition matrix must be changed, and these are dependant on the reliability of the equipment and its subsystems, the maintainability of the equipment and the maintenance policies of the operators of the equipment. So Markov analysis can be used to study management actions as well as engineering ones.

6.2.3 The continuous case

In this case, the system states will be able to change from moment to moment, rather than from day to day, and so it is necessary to consider failure and repair rates, rather than failure and repair probabilities. Consider the following example. An Auxiliary Power Unit (APU) in a military vehicle consists of two generators. If one fails, it can be repaired by the crew, but if they both fail (i.e. the survivor fails before the failed one is repaired), then REME (the maintenance engineers of the British army) will come out and replace the whole unit. This can be represented by the diagram in Fig. 6.3.

It is necessary to quantify this situation, and this is shown in Fig. 6.4, where

λ is the failure rate of one generator when two are

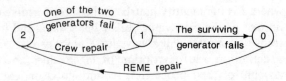

Fig. 6.3 Preliminary state space diagram of APU.

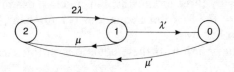

Fig. 6.4 Quantitative state space diagram of APU.

Table 6.2 Table of rates of transition

		Now		
		2	1	0
	2		μ	μ'
Later	1	2λ		0
	0	0	λ'	

functioning

λ' is the failure rate of one generator when only one is functioning

μ is the crew repair rate

μ' is the REME repair rate.

This information can be summarized in Table 6.2.

Note that this is a table of rates, and not of probabilities, as is the transition matrix in the discrete case. The diagonals are left blank, as there is no such thing as a rate at which something remains unchanged. The zeros mean that it is impossible to go straight from state 2 to state 0 and from state 0 to state 1 in this case, and this is reflected in the state space diagram by the fact that there are no arrows going to and from those particular states. Remember that failure rate has the property that $\lambda \delta t$ is the probability that an item that is functioning at time t fails in the interval $(t, t + \delta t)$. Similarly, for repair rate μ, μ has the property that the probability of an item that is in the failed state at time t being in the functioning state by time $t + \delta t$ is $\mu \delta t$. (Note that the MTTR is given by $1/\mu$ for a constant repair rate μ.) For these reasons it is necessary to use calculus to obtain a transition matrix, as shown in Table 6.3.

Table 6.3 Table of transition probabilities

		State at time t		
		2	1	0
State	2	$1 - 2\lambda\delta t$	$\mu\delta t$	$\mu'\delta t$
at time	1	$2\lambda\delta t$	$1 - (\mu + \lambda')\delta t$	0
$t + \delta t$	0	0	$\lambda'\delta t$	$1 - \mu'\delta t$

The following matrix, which is obtained from the tables, is called the rate transition matrix:

$$\begin{array}{ccc} -2\lambda & \mu & \mu' \\ 2\lambda & -(\lambda' + \mu) & 0 \\ 0 & \lambda' & -\mu' \end{array}$$

In order to obtain the availability of the system, it is necessary to solve some differential equations. Suppose $A_i(t)$ is the probability that the system is in state i at time t, $i = 1, 2, 3$. Then the A_i are the solutions of the following differential equations:

$$\frac{dA_2}{dt} = -2\lambda A_2 \quad + \mu A_1 + \mu' A_0$$

$$\frac{dA_1}{dt} = 2\lambda A_2 - (\mu + \lambda') A_1$$

$$\frac{dA_0}{dt} = \quad\quad \lambda' A_1 - \mu' A_0.$$

The derivation of these equations is as follows. Consider Table 6.4, obtained from a general state transition matrix, where $\lambda_{i,j}$ is the rate at which the system changes from state j to state i.

Table 6.4 General table of transition probabilities

		State at time t		
		2	1	0
State	2	$1 - (\lambda_{21} + \lambda_{31})\delta t$	$\lambda_{12}\delta t$	$\lambda_{13}\delta t$
at time	1	$\lambda_{21}\delta t$	$1 - (\lambda_{12} + \lambda_{32})\delta t$	$\lambda_{23}\delta t$
$t + \delta t$	0	$\lambda_{31}\delta t$	$\lambda_{32}\delta t$	$1 - (\lambda_{13} + \lambda_{23})\delta t$

If $A_i(t)$ is the probability of being in state i at time t, then as before, considering the case when $j = 1$,

$$A_1(t + \delta t) = A_1(t)[1 - (\lambda_{21} + \lambda_{31})]\delta t + A_2(t)\lambda_{21}\delta t + A_3(t)\lambda_{31}\delta t.$$

The relationship leading to this equation is illustrated in diagrammatic form in Fig. 6.5.

Now, rearranging, to collect all the terms multiplied by δt on the right-hand side,

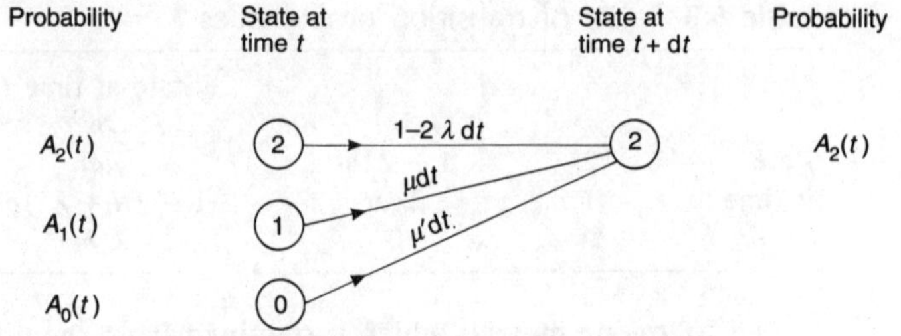
Probability
State at time t
State at time t + dt
Probability
A2(t)
A1(t)
A0(t)
2
1
0
1–2 λ dt
μdt
μ'dt
2
A2(t)

Fig. 6.5 Derivation of differential equations.

$$A_1(t+\delta t) - A_1(t) = -A_1(t)(\lambda_{21}+\lambda_{31})\delta t + A_2(t)\lambda_{21}\delta t + A_3(t)\lambda_{31}\delta t.$$

Now divide both sides of the equation by δt

$$\frac{A_1(t+\delta t) - A_1(t)}{\delta t} = -A_1(t)(\lambda_{21}+\lambda_{31}) + A_2(t)\lambda_{21} + A_3(t)\lambda_{31}.$$

As $\delta t \to 0$, then the left–hand side of the above equation tends to the derivative, giving, in the limit

$$\frac{dA_1}{dt} = -A_1(t)(\lambda_{21}+\lambda_{31}) + A_2(t)\lambda_{21} + A_3(t)\lambda_{31}.$$

The other differential equations of the system follow similarly. Notice how the coefficients of the equations are derived from the rate transition matrix. The techniques for solving equations of this nature are not important here, although notes on Laplace transforms are given in a later section. See, for example, Grosh (1989) for further discussion on this point. There is not a closed form for the solution, but in the special case that

$$\lambda = 0.01\ \text{fhr}^{-1}$$

$$\lambda' = 0.015\ \text{fhr}^{-1}$$

$$\mu = 2.0\ \text{rhr}^{-1}$$

$$\mu' = 3.0\ \text{rhr}^{-1}$$

and assuming the initial condition

$$A_2(0) = 1$$

$$A_1(0) = 0$$

$$A_0(0) = 0$$

i.e. it is certain that the system is working perfectly at the start, then the solution to the system is

$$A_2 = \quad 0.9901 + 9.98 \times 10^{-3} \exp(-2.0353t) - 1.06 \times 10^{-4} \exp(-2.997t)$$

$$A_1 = 9.83 \times 10^{-3} - 9.83 \times 10^{-3} \exp(-2.0353t) - 2.15 \times 10^{-6} \exp(-2.997t)$$

$$A_0 = 4.91 \times 10^{-5} - 1.53 \times 10^{-4} \exp(-2.0353t) + 1.04 \times 10^{-4} \exp(-2.997t).$$

The important thing to note is the form of this solution, which is typical of the form of solutions to all Markov problems. Each of the As is the sum of three terms, one of which is a constant, and the other two are negative exponentials. Notice also, that the solutions sum to one, as the As are probabilities, and that putting $t = 0$ gives the initial conditions, as indeed it should.

In this case the A_i corresponds to availability, and the long term, or steady state availabilities are obtained by letting t tend to infinity, in which case the negative exponentials tend to zero, giving

$$A_2 = 0.9901$$

$$A_1 = 9.83 \times 10^{-3}$$

$$A_0 = 4.91 \times 10^{-5}$$

in the long term. If either of states 2 or 1 are acceptable, then the systems availability is given by

$$A_2 + A_1 = 99.99\%.$$

This last result can be obtained by observing that as t tends to infinity, then the derivatives of the A_i tend to zero. If the derivatives in the differential equations above are put to zero, then the A_i are the solutions of the three linear equations

$$0 = -2\lambda A_2 \qquad\qquad + \mu A_1 + \mu' A_0$$

$$0 = \quad 2\lambda A_2 - (\mu + \lambda') A_1$$

$$0 = \qquad\qquad\qquad \lambda' A_1 - \mu' A_0.$$

To which must be added the equation

$$1 = A_2 + A_1 + A_0$$

for the same reasons that the same equation was added in the discrete case. The solutions to these equations are given above, for the stated values of λ, λ', μ, μ'.

The graphs of the As as functions of time are shown in Fig. 6.6. Note that each A_i rises or falls exponentially to its steady state value. In Fig. 6.7 is shown a graph of a typical life of such an auxiliary power unit (APU). The steady state values of the As are simply the ratios of the times the APU spends in each state to the total time, taken over a long time period. The curves shown in Fig. 6.6 are the average values, taken over a large number of vehicles, of actual cases as illustrated in Fig. 6.7.

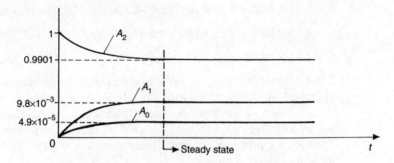

Fig. 6.6 Graph showing A_i as a function of time.

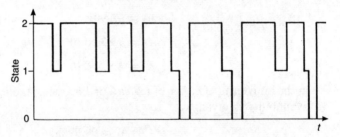

Fig. 6.7 Typical history of APU.

The availability of an equipment is the ratio of uptime to total time. The reliability gives a measure of the length of time the equipment stays functioning, or the length of just one of the functioning cycles of Fig. 6.7, as shown in Fig. 6.8, or the length of time before REME have to be called out. The state space diagram for this situation is shown in Fig. 6.9, and note that it is the same as Fig. 6.4, but without the REME repair. Note too that once the system is in S_0, it cannot get out of it. S_0 is called an absorbing state. From Fig. 6.9, the rate transition matrix for the system is:

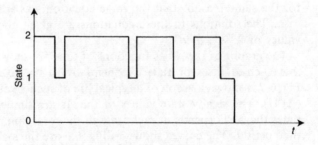

Fig. 6.8 Typical single cycle of APU.

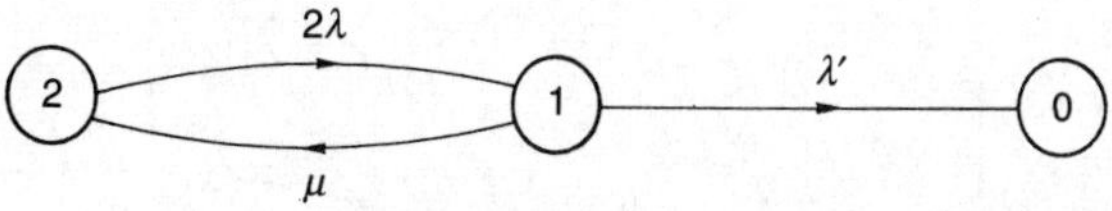

Fig. 6.9 State space diagram: reliability of APU.

$$
\begin{array}{ccc}
-2\lambda & \mu & 0 \\
2\lambda & -(\lambda' + \mu) & 0 \\
0 & \lambda' & 0
\end{array}
$$

If $R_i(t)$ is the probability the system is state S_i at time t, then the differential equations for the R_i are

$$\frac{dR_2}{dt} = -2\lambda R_2 + \mu R_1$$

$$\frac{dR_1}{dt} = 2\lambda R_2 - (\mu + \lambda')R_1$$

$$\frac{dR_0}{dt} = \lambda' R_1$$

The solution to this system is

$$R_2 = 0.9902\exp(-1.5\times10^{-3}t) + 9.8\times10^{-3}\exp(-2.03t)$$
$$R_1 = 9.8\times10^{-3}\exp(-1.5\times10^{-3}t) - 9.8\times10^{-3}\exp(-2.03t)$$
$$R_0 = -1.001\exp(-1.5\times10^{-3}t) + 7.2\times10^{-5}\exp(-2.03t) + 1.0$$

and the graphs of the Rs as functions of time are shown in Fig. 6.10. Note that the form of these equations is just as described earlier, and that as t tends to infinity, R_0 tends to one, indicating that if you wait long enough, the system is sure to fail eventually.

In this case, either of the first two states is acceptable, and so the reliability, R, is given by

$$R = R_2 + R_1$$

$$= 1.001\exp(-1.5\times10^{-3}t) - 7.2\times10^{-5}\exp(-2.03t)$$

the graph of which is shown in Fig. 6.11.

Once the reliability is known, the elementary relationships between $f(t)$, MTBF and $R(t)$ can be used to obtain more information.

$$f(t) = -\frac{dR}{dt}$$

$$= 1.5\times10^{-3}\exp(-1.5\times10^{-3}t) - 1.5\times10^{-4}\exp(-2.03t)$$

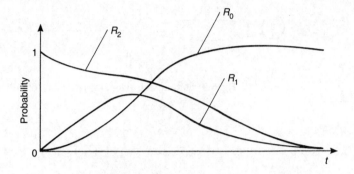

Fig. 6.10 Graph of R_i as a function of time.

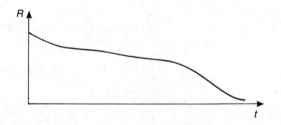

Fig. 6.11 Graph of reliability of the APU as a function of time.

and

$$\text{MTBF} = \int_0^\infty R(t)\,\mathrm{d}t$$

$$= 6000\,\text{hr}.$$

6.2.4 Modelling examples

The objective of this section is not to acquaint the students with differential equations or the means to solve them, but to improve their modelling capabilities. Markov processes are useful for modelling a number of situations that are difficult if not impossible to model by other techniques. A number of examples of models of different types of systems are presented in order to make the point. Grosh (1989) and Pagés and Gondran (1986) have further examples of Markov models.

Examples 6.1

6.1.1 The example considered so far was of a system with the capacity to operate when in a degraded state, but there were two types of repair policies in operation, namely repair from the degraded state when applicable (crew repair), and repair to the fully operational state once the

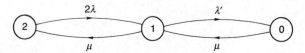

Fig. 6.12 State space diagram: system with one repair team.

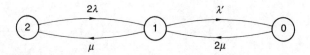

Fig. 6.13 State space diagram: system with two repair teams.

system had failed (REME repair). In a different situation, the system could consist of two units, but each is repaired separately, so there is the possibility of repairing from the failed state to the degraded state, as illustrated in Figs 6.12 and 6.13.

An example of this type of system could be a system of two generators on board a ship. They would be repaired separately, and in the case that they are both down together, first one is repaired, and brought into operation, and then the second is repaired. If there is only one repair team, and only one generator can be repaired at a time, Fig. 6.12 is applicable, while if there are two repair teams, and they can both be worked on at the same time, then the situation is described in Fig. 6.13.

In this case the steady state availabilities can be found without much difficulty, and it turns out that for the first situation, only one repair team then

$$A = \frac{\mu^2 + 2\lambda\mu}{\mu^2 + 2\lambda\mu + 2\lambda^2}$$

and when there are two repair teams, then

$$A = \frac{\mu^2 + 2\lambda\mu}{\mu^2 + 2\lambda\mu + \lambda^2}.$$

Note that in the second case, the unavailability is approximately halved, if μ is very much greater than λ.

6.1.2 In the case that only one of the two units is operated at any time (standby redundancy), then of course the failure rates are different, and Fig. 6.14 shows this situation, where λ is the failure rate of an operational unit, while λ' is the dormant failure rate.

This model assumes that the two units are identical, but the case that they are different, and have different failure rates can easily be catered for. Fig. 6.14 does not show the maintenance, which depends on the

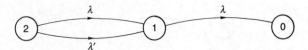

Fig. 6.14 State space diagram: standby redundancy with dormant failure. Maintenance not shown.

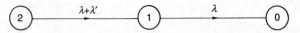

Fig. 6.15 State space diagram: more usual representation of Fig. 6.14.

policy chosen and the situation etc., and the comments made earlier apply. It is not usual to leave a state space diagram with two arrows with the same tail and head, and the situation described by Fig. 6.14 would be drawn as shown in Fig. 6.15.

Figure 6.15 shows the situation when there is frequent inspection of the dormant unit to determine its state. If, on the other hand, the standby item is never inspected, then state 1 must be divided into two sub-states, as shown in Fig. 6.16. One is the state in which the operating item has failed, and is possibly being repaired, but the operator is aware of the situation, while the other sub-state, S_1' is the state in which the dormant item has failed, but the operator is unaware of it.

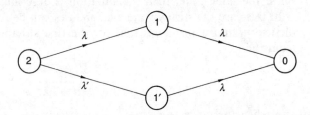

Fig. 6.16 State space diagram: standby redundancy with no inspection.

6.1.3   In the standby case, there is also the possibility of a switching failure. Suppose that s is the probability that the dormant item will not be able to be brought onto line, for whatever reason. The state space diagram for this situation is shown in Fig. 6.17.

6.1.4   The system may have more than one failed state, or it may be necessary to model it with more than one failed state. Considering the APU example again, suppose consideration of the engine is added to the model, then the state space diagram is that shown in Fig. 6.18, where the second number in each circle shows the number of generators functioning, as before, while the 1 shows the states where the engine is

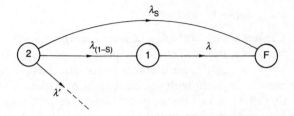

Fig. 6.17 State space diagram: switching failure.

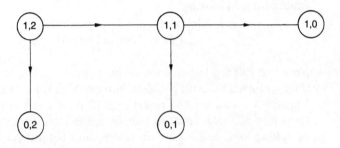

Fig. 6.18 State space diagram: APU including engine.

working, and the 0 shows when it is failed, and λ_m, λ_g are the failure rates of the engine and generators respectively. Note that no maintenance is shown on this diagram.

With any modelling technique there are problems, and this technique is no exception. The first is the complexity of the models, and the difficulties of solving the differential equations. There is no simple solution to this, although computers have made great inroads into the problem of obtaining numerical solutions to the equations. In particular, the computing language APL, which deals with matrices, and operates in a matrix language, has the ability to compute the eigenvalues and eigenvectors of a square matrix, which are needed to solve the differential equations.

The second difficulty is the one associated with the constant failure and repair rates. If the rates are constant, then there is no change in the system until it actually changes state. Put another way, a constant failure rate assumes the system does not age until the instant of failure, and similarly a constant repair rate assumes no improvement until the moment of repair, and the unit is back on line. This last is clearly unrealistic. There is a technique for dealing with this problem, but it has the effect of introducing more states, and so making the model more complex, and will not be dealt with here.

6.2.5 Laplace transforms

Laplace transforms are a means of solving the differential equations that were met above. They transform the differential equation into a relatively straightforward linear equation. Here only an outline of the method is given, with just enough information to solve the type of equation that was met earlier in the section, and the means to find the solution by use of look-up tables. For a more detailed study of the technique, a good textbook on calculus for engineers would be suitable.

The Laplace transform of the function $f(t)$, $L(f)$ is defined to be the function $\bar{f}(s)$ given by:

$$L(f) = \bar{f}(s) = \int_0^\infty f(t)e^{-st}dt.$$

Note that this is a transformation on a function. The idea of transformations to turn a difficult problem into one that is less difficult is not new. Logarithms were used as recently as 20 years ago to turn multiplication sums into addition ones. In that case, the transformation was obtained by taking logs, using look-up tables, and the answer was obtained by transforming the result back again, also by the use of look-up tables. The big difference between logs and Laplace transforms is that logs are a transformation on numbers, while Laplace transforms functions.

It is easy to show by doing the integration that if

$$f(t) = e^{-\lambda t}$$

then

$$L(f) = \bar{f}(s) = \frac{1}{s + \lambda}$$

and other results can be obtained similarly. All the results we shall need are contained in Table 6.5.

Examples 6.2   **6.2.1** Consider a system consisting of only two states, functioning and failed, called S_1 and S_2 respectively. Suppose the system starts in the functioning state, and the state space matrix is

$$-\lambda \quad 0$$
$$\lambda \quad 0$$

i.e. we are looking at the reliability of a simple two state system with failure rate λ, and no repair. As the solution to this problem is known, it is a check on the Laplace transform method in this case.

If $R(t)$ is the probability of being in state 1 and $F(t)$ the probability of being in state 2 at time t, then the differential equations for R and F are

Table 6.5 Laplace transforms

$f(t)$	$L(f) = \bar{f}(s)$
1	$1/s$
$e^{-\lambda t}$	$\dfrac{1}{s + \lambda}$
$\dfrac{\mathrm{d}f}{\mathrm{d}t}$	$-f(0) + s\bar{f}(s)$
$af(t) + bg(t)$ a, b constant	$a\bar{f}(s) + b\bar{g}(s)$

$$\dot{R} = -\lambda R$$

$$\dot{F} = \lambda R$$

and $R(0) = 1, F(0) = 0$. That is, the system is certain to be in the functioning state at time $t = 0$. Taking Laplace transforms gives

$$L(\dot{R}) = -\lambda L(R)$$

$$L(\dot{F}) = \lambda L(R)$$

or, using the results in Table 6.5,

$$-1 + s\bar{R}(s) = -\lambda \bar{R}(s)$$

$$s\bar{F}(s) = \lambda \bar{R}(s)$$

which, by rearranging the first equation to obtain $\bar{R}$ and then substituting in the second to give $\bar{F}$, gives the solutions:

$$\bar{R} = \frac{1}{s + \lambda},$$

$$\bar{F} = \frac{\lambda}{s(s + \lambda)} = \frac{1}{s} - \frac{1}{s + \lambda}$$

the last result by the use of partial fractions, which inverting by use of Table 6.5, gives

$$R = e^{-\lambda t}$$

$$F = 1 - e^{-\lambda t}$$

which confirms a result we already knew.

6.2.2 Suppose we add repair to the last example, so that the rate transition matrix becomes

$$\begin{matrix} -\lambda & \mu \\ \lambda & -\mu \end{matrix}$$

where μ is the repair rate. Then the differential equations are now:

$$\dot{R} = -\lambda R + \mu F$$
$$\dot{F} = \lambda R - \mu F$$

and taking the Laplace transform, this becomes,

$$-1 + s\bar{R} = -\lambda\bar{R} + \mu\bar{F}$$
$$s\bar{F} = \lambda\bar{R} - \mu\bar{F}$$

which has solution

$$\bar{R} = \frac{s + \mu}{s(s + \lambda + \mu)} = \frac{\mu}{s(\lambda + \mu)} + \frac{\lambda}{(s + \lambda + \mu)(\lambda + \mu)}$$

$$\bar{F} = \frac{\lambda}{s(s + \lambda + \mu)} = \frac{\lambda}{s(\lambda + \mu)} - \frac{\lambda}{(s + \lambda + \mu)(\lambda + \mu)}$$

after applying the technique of partial fractions, which gives

$$R = \frac{\mu}{\lambda + \mu} + \frac{e^{-\lambda t}}{\lambda + \mu}$$

$$F = \frac{\lambda}{\lambda + \mu} - \frac{e^{-\lambda t}}{\lambda + \mu}$$

which as $t \to \infty$, and substituting for λ and μ, gives the well known result

$$R = \frac{\mu}{\lambda + \mu} = \frac{\text{MTBF}}{\text{MTBF} + \text{MTTR}}$$

$$F = \frac{\lambda}{\lambda + \mu} = \frac{\text{MTTR}}{\text{MTBF} + \text{MTTR}}.$$

These simple results, much of which was already known, illustrate the method and give confidence that it works. Of course, it is always possible to verify the proposed solution to a differential equation by differentiating and substituting in the initial equation! One more example will be presented to illustrate the method.

6.2.3 Consider the system illustrated in Fig. 6.13. This has rate transition matrix

$$
\begin{matrix}
-2\lambda & \mu & 0 \\
2\lambda & -(\lambda + \mu) & 2\mu \\
0 & \lambda & -2\mu
\end{matrix}
$$

which gives the differential equations

$$\dot{P_2} = -2\lambda P_2 + \mu P_1$$
$$\dot{P_1} = 2\lambda P_2 - (\lambda + \mu)P_1 + 2\mu P_0$$
$$\dot{P_0} = \lambda P_1 - 2\mu P_0.$$

Assuming the system is certain to be in the fully functioning state at the start, i.e. $P_2(0) = 1$ while $P_0(0) = P_1(0) = 0$, then taking the Laplace transform of the equations gives

$$-1 + s\bar{P}_2 = -2\lambda\bar{P}_2 + \qquad \mu\bar{P}_1$$

$$s\bar{P}_1 = \qquad 2\lambda\bar{P}_2 - (\lambda + \mu)\bar{P}_1 + 2\mu\bar{P}_0$$

$$s\bar{P}_0 = \qquad\qquad\qquad\qquad \lambda\bar{P}_1 - 2\mu\bar{P}_0$$

which solving for the P_i gives

$$\bar{P}_2 = \frac{s^2 + s(\lambda + 3\mu) + 2\mu^2}{\Delta}$$

$$\bar{P}_1 = \frac{2\lambda(s + 2\mu)}{\Delta}$$

$$\bar{P}_0 = \frac{2\lambda^2}{\Delta}$$

where

$$\Delta = s(s + \lambda + \mu)(s + 2(\lambda + \mu))$$

and applying partial fractions, this gives

$$\bar{P}_2 = \frac{\mu^2}{s(\lambda + \mu)^2} + \frac{2\lambda\mu}{(s + \lambda + \mu)(\lambda + \mu)^2} + \frac{\lambda^2}{(s + 2(\lambda + \mu))(\lambda + \mu)^2}$$

$$\bar{P}_1 = \frac{2\lambda\mu}{s(\lambda + \mu)^2} + \frac{2\lambda(\lambda - \mu)}{(s + \lambda + \mu)(\lambda + \mu)^2} + \frac{2\lambda^2}{(s + 2(\lambda + \mu))(\lambda + \mu)^2}$$

$$\bar{P}_0 = \frac{\lambda^2}{s(\lambda + \mu)^2} - \frac{2\lambda^2}{(s + \lambda + \mu)(\lambda + \mu)^2} + \frac{\lambda^2}{(s + 2(\lambda + \mu))(\lambda + \mu)^2}$$

which, inverting the Laplace transform by the use of Table 6.5, gives the following for the P_i,

$$P_2 = \frac{\mu^2 + 2\lambda\mu e^{-(\lambda + \mu)t} + \lambda^2 e^{-2(\lambda + \mu)t}}{(\lambda + \mu)^2}$$

$$P_1 = \frac{2\lambda\mu + 2\lambda(\lambda - \mu)e^{-(\lambda + \mu)t} - 2\lambda^2 e^{-2(\lambda + \mu)t}}{(\lambda + \mu)^2}$$

$$P_0 = \frac{\lambda^2 - 2\lambda^2 e^{-(\lambda + \mu)t} + \lambda^2 e^{-2(\lambda + \mu)t}}{(\lambda + \mu)^2}.$$

Exercises 6.1

6.1.1 Consider Example 6.1.1 in the main text and Figs 6.12 and 6.13. Write down the rate transition matrices, and hence obtain the availabilities stated.

6.1.2 Consider Example 6.1.4 in the text, and Fig. 6.18 illustrating it, which does not show any maintenance. Redraw Fig. 6.18 showing maintenance, modelling the following repair policies:

(a) Repair subsystems as they fail, but only one repair team that can deal with any subsystem.

(b) As (a), but two specialist repair teams, one for each of the two subsystems. The teams cannot repair the subsystem which is not their specialty.

What other policies can you think of? Model them too (there are at least two).

6.1.3 Considering Example 6.1.4 Fig. 6.18 again, as shown it models the situation in which the surviving subsystems are not left running once the system is considered to be failed. Model the situation in which they are left running when the system has failed (leaving out repair).

6.1.4 The subsystem of a car consisting of the four road wheels and the spare is considered to be functioning if any four of the wheels are functioning and in place. Draw the state space diagram for this system.

6.1.5 A series system consists of two identical components with constant failure and repair rates λ and μ respectively. Show that the availability of each is given by

$$A = \frac{\mu}{\lambda + \mu}$$

and hence by drawing an RBD find the availability of the system.

The system can also be modelled by either of the two state space diagrams shown in Figs 6.12 and 6.13 in the text, if S_2 is considered to be the functioning state, and S_0 and S_1 the failed states, as well as the system shown in Fig. 6.19. Show that the long term availability of this last system is given by

$$A = \frac{\mu}{\mu + 2\lambda}.$$

This gives four ways of modelling the same system, and three different values for the long term availability. Why the differences? Explain.

6.1.6 An industrial process plant consists of four trains fed by five compressors, arranged geometrically as shown in Fig. 6.20. Given that:

(1) all four trains are needed for the system to function;

(2) the lines joining the compressors and trains indicate which compressors can serve which trains;

(3) each compressor can serve only one train at a time;

(4) a compressor and train take approximately the same time to repair;
(5) the unavailabilities are:
   (i) compressors, 2×10^{-3};
   (ii) trains, 7×10^{-5}.

then:
   (a) Draw the RBD for the system.
   (b) Draw the state space diagram for the system.
   (c) Write down a rate transition matrix for the system.

What assumptions are you making about repair and maintenance? What alternative assumptions could you make? Would it make any difference?

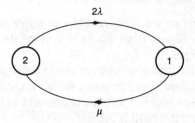

Fig. 6.19 State space diagram: a system of two components in series that stops on failure.

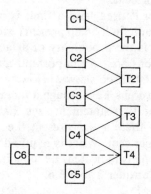

Fig. 6.20 Diagrammatic representation of trains and compressors.

6.3 Bayesian statistics

6.3.1 Introduction

As an alternative to the classical, frequentist approach, the use of Bayesian statistics is not uncommon in reliability analysis and modelling.

It has two advantages — it allows the input of subjective data (engineering judgment), and probability is defined in a broader way that extends its use to a measure of uncertainty, which is less restricted and more in line with common usage than the classical, frequentist approach. It depends upon Bayes' theorem, which is a theorem in classical statistics, which is presented in Section 6.3.2. The alternative use of probability is explained in Section 6.3.3, and Bayesian analysis in Section 6.3.5. There are few textbooks on Bayesian statistics compared with the number on classical statistics, but Berger (1985) is one, and Sander and Badoux (1991) is the result of a course on Bayesian methods in reliability.

6.3.2 Bayes' theorem

This theorem was discovered by an eighteenth century vicar, the Reverend Thomas Bayes, from whom it derived its name. It is best illustrated by an example.

Let us suppose a supplier is advertising a diagnostic tool for a piece of equipment. Further suppose that it consists of an electronic black box that can be plugged into the equipment of interest, and that there is some indication if a fault is present — a light flashes or a bell rings or something similar. He claims that if there is a fault present, then there is a 99 per cent chance that it is indicated. On the other hand, there is a 2 per cent chance of a false alarm, i.e. a fault being indicated when in fact there is not one present.

If the tool is used, and a fault is indicated, then what is the actual probability of a fault being present? More to the point, if no fault is indicated, what is the probability of a fault being present? The answers to these questions are very important, and they are not 99 per cent and 2 per cent, as shall be shown below.

The confusion arises through a failure to understand conditional probabilities. The first statement says that the probability of a fault being indicated, if there is one present, i.e. conditional on the presence of a fault, is 99 per cent. What is required is the probability of a fault being present, if one is indicated, i.e. conditional on indication, and these are different. Consider Table 6.6.

Table 6.6 Joint probabilities of absence, or presence, of a fault and absence, or presence, of indication

| | | Fault | | |
		Present	Absent	Totals
Indication	Yes	0.00495	0.0199	0.02485
	No	0.00005	0.9751	0.97515
Totals		0.005	0.995	1.0

Looking at the totals, 0.005 is the probability of a fault being present when the equipment is tested, and 0.995 is the probability of it being fault free. Of the 0.5 per cent of the time that a fault is present, 99 per cent of the time it will be indicated. Since 99 per cent of 0.005 is 0.004 95, this is the probability of a fault being present and it being indicated. Note that 1 per cent of faults remains undetected, and 1 per cent of 0.005 is 0.000 05, which is shown in the table. Of the 99.5 per cent of the time that there is no fault, 2 per cent of the time there will be a false alarm, i.e. a fault will be indicated. Since 2 per cent of 99.5 per cent is 0.0199, this is the probability of a fault being indicated, even though no fault is present. Also 98 per cent of 99.5 per cent is 0.9751, and this is the probability of no fault being shown when there is no fault present. Summing along the rows, there will be a 0.024 85 probability of a fault being indicated overall, and a 0.975 15 probability of no fault being indicated.

We used the probabilities of indication, and the overall probability of a fault being present to fill in the table, and then find the probability, overall, of indication. This process can now be reversed to find the probability of a fault being present, if the bell rings. This involves finding what 0.004 95 is as a proportion of 0.024 85, i.e. the probability of simultaneously a fault being present and a fault being indicated as a proportion of the probability of the bell ringing, and this is

$$0.004\,95/0.024\,85 = 0.199$$

or nearly 20 per cent. What we have just done is to derive Bayes' theorem. This is explained below.

If F is used to mean 'a fault is present', and I to mean 'A fault is indicated', then these two different probabilities are written

$$P(I|F) = \text{the probability of I given F}$$

$$P(F|I) = \text{the probability of F given I.}$$

$P(F)$ is the unconditional probability of a fault, and is called the marginal. We were given $P(I|F)$ and $P(I|\bar{F})$, where ‾ indicates converse. Table 6.6 is rewritten in Table 6.7 in terms of the marginal and conditional probabilities that we know. This is just putting into mathematical symbols the descriptive paragraph above.

What we require is $P(F|I)$, which is given by

$$P(F|I) = \frac{P(I|F)P(F)}{P(I)}$$

and

$$P(I) = P(I|F)P(F) + P(I|\bar{F})P(\bar{F})$$

and this is Bayes' theorem.

This simple example shows how Bayes' theorem can be used in practice.

Table 6.7 Joint probabilities of absence, or presence, of a fault and absence, or presence, of indication, in symbols

		Fault		
		Present	Absent	Totals
Indication	Yes	$P(F)P(I\|F)$	$P(\bar{F})P(I\|\bar{F})$	$P(F)P(I\|F) + P(\bar{F})P(I\|\bar{F})$
	No	$P(F)P(\bar{I}\|F)$	$P(\bar{F})P(\bar{I}\|\bar{F})$	$P(F)P(\bar{I}\|F) + P(\bar{F})P(\bar{I}\|F)$
Totals		$P(F)$	$P(\bar{F})$	1.0

Notice how the theorem works. It reverses the conditional probabilities, but needs a marginal probability too. It is often written as a proportion, as follows

$$P(F|I) \propto P(I|F)P(F).$$

As an aside, it is interesting to note just how low the efficiency of the system is at 20 per cent! Management, and the mechanics on the ground, would not be impressed. It is left as an exercise for the reader to calculate the probability that a fault is undetected in the event of there being no indication. It is also interesting to see just what can be done to improve the situation. It is no good increasing the probability of detecting a fault when one is present—that is already at 99 per cent, and examining the structure of the above calculations, putting it up to 100 per cent would not improve matters much. It is the 2 per cent false alarm rate that has to be decreased, and to a figure far below 0.5 per cent, which is the failure rate. The probability of a false alarm for a fire alarm must be an order of magnitude lower than the probability of a fire, otherwise too many alarms will be false!

A further example will illustrate the theorem. This example is known as the prosecutor's dilemma, and although not of interest in its present form to reliability engineers, is topical at the time of writing.

In court, forensic evidence against the accused is sometimes a DNA match. The claim is that as the probability of a match is very low, the order of one in a million, the probability of the accused being innocent is also very low. This is not necessarily the case, as Bayes' theorem shows.

The probability of 10^{-6} is the conditional probability of there being a match if the accused is innocent. This is essentially a significance test, with the null hypothesis being that the accused is innocent. What is of interest is the conditional probability of the accused being innocent if there is a match. Let G be guilt, and M be match. Then

$$P(M|\bar{G}) = 10^{-6}$$

and

$$P(M|G) = 1$$

i.e. if the accused is guilty there is certain to be a match. Then

$$P(G|M) \propto P(M|G)P(G).$$

The marginal $P(G)$ can be tricky, but suppose there is no further evidence, and the crime took place in a large city where there are 50 000 potential suspects, then

$$P(G) = \frac{1}{50\,000}$$

which gives

$$P(G|M) \propto 1 \times \frac{1}{50\,000} = 2 \times 10^{-5}$$

and

$$P(\bar{G}|M) \propto P(M|\bar{G})P(\bar{G})$$

$$1 \times 10^{-6} \times \frac{49\,999}{50\,000} = 10^{-6}.$$

$P(G|M)$ and $P(\bar{G}|M)$ must sum to one, and as the constant of proportionality is the same in both cases, normalizing so that the sum is one gives:

$$P(\bar{G}|M) = 4.8 \text{ per cent}$$

which although small, is a far cry from one in a million. Incidentally, this example demonstrates that the significance is not the probability that the null hypothesis is true, and in fact is far from it. As can be seen, the probability of the truth of the null hypothesis is not easy to calculate, and involves the need for more data or assumptions.

6.3.3 Bivariate distributions

The statistics studied so far assumes that only one variable, usually the time to or between failure(s), is being studied. The classical use of Bayes' theorem is when two (or more) variables have to be considered at the same time. In the above example, the two variables were the presence/absence of a fault and the presence/absence of an alarm. In this case they could also only take one of two values. In practice, the variables concerned usually can take more than two values, and they are often continuous. To understand Bayes' theorem, it is necessary to have some idea of bivariate distributions, and the object of this section is to explain what bivariate distributions are, and some of the terms used in their study.

When studying one variable, like time to failure, the idea of a histogram is well understood, and the way the probability density function

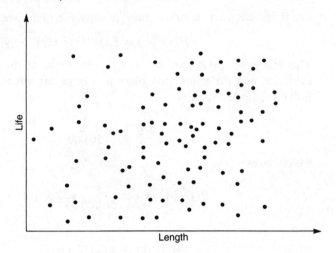

Fig. 6.21 Scatter diagram of length and lives of crankshafts.

can be derived from the histogram, by taking more and more data, and making the interval smaller and smaller, although not possible in practice, is generally appreciated. It may be necessary to study two variables at once — suppose we wished to understand how the length of a crankshaft influenced its reliability, we could, in principle, measure the lengths of a large number of crankshafts, and then test them all to failure, recording the life of each. Our data now consist of pairs of readings, one pair for each crankshaft, each pair consisting of the length and life of one crankshaft. Each pair can be represented by a point on a plane, by considering each pair to be the coordinates of a point, with say, the X-axis representing the length and the Y-axis the life. This is illustrated in Fig. 6.21.

In order to construct a histogram of these data, they first have to be grouped, just as was done when analysing a single variable. The two axes are divided up into intervals in the same way that they would be if a single variable were being analysed. This divides the plane into rectangles, and in each rectangle there are a number (which may be zero) of points. The (i, j)th rectangle, that is the one corresponding to the ith interval along the X-axis and the jth interval along the Y-axis, has $f_{i,j}$ points in it. This is shown in Fig. 6.22, which shows the points and the rectangles, and Fig. 6.23, which shows the values of $f_{i,j}$. To draw the histogram requires three dimensional paper, as the histogram consists of the rectangles shown in Fig. 6.23 with pillars with heights proportional to the values of $f_{i,j}$ above each rectangle. A sketch of such a histogram is shown in Fig. 6.24.

Just as with the single variable case, if we were able to collect a lot

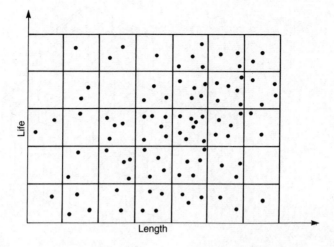

Fig. 6.22 Scatter diagram of length and lives of crankshafts divided into regions.

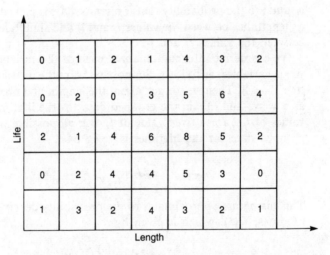

Fig. 6.23 Length and lives of crankshafts showing frequencies.

of data, and make the interval lengths small, then the histogram would approximate more and more closely to a smooth surface, $f(l, t)$, where l is the length and t is the lifetime. This is the joint pdf of l and t. Probabilities are now given by volumes under this surface, and volumes are obtained by evaluating double integrals, so that, for example, the integral:

$$\int_{l_1}^{l_2} \int_{t_1}^{t_2} f(l, t) \, \mathrm{d}l \, \mathrm{d}t$$

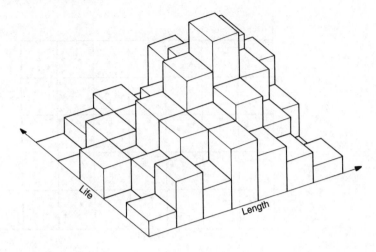

Fig. 6.24 Histogram of length and lives of crankshafts.

is equal to the probability that if a crankshaft is picked at random, then its length lies between the values l_1 and l_2 and simultaneously its life lies between the values t_1 and t_2.

We can derive information about each of the variables, considered as single variables, as follows. Suppose we want the distribution of lifetimes (the Y-axis). Then we 'forget' about the length, and project horizontally, as it were, and obtain the grouped data for the lifetimes by adding the values of $f_{i,j}$, for a fixed value of j, over all possible values of j. So the grouped data for the lifetimes is given by

$$f_{*,j} = \sum_i f_{i,j}.$$

This histogram approximates to the probability density function for the lifetimes, $f_2(t)$, which is given by

$$f_2(t) = \int f(l, t)\mathrm{d}l$$

where the integral is taken over all values of l. This is known as the marginal distribution of t. Similarly the marginal distribution of l is given by

$$f_1(l) = \int f(l, t)\mathrm{d}t.$$

The marginal distribution gives the distribution of a single variable, ignoring any effects of the other variable(s).

Going back to the data, if we fix i at a given value, and consider the $f_{i,j}$, we get a histogram which will, in general depend on the value of

i chosen. This is the distribution of lifetimes for a particular length, and is the conditional distribution. If we go over to the probability density function, then if l is fixed, t has a pdf given by

$$f(t|1) \propto f(l, t), \quad l \text{ fixed.}$$

Note that it is only proportional. The area under the curve must be unity, and so we must divide by the area, which is given by integrating with respect to t,

$$f(t|l) = \frac{f(l, t)}{\int f(l, t) \, \mathrm{d}t} \quad l \text{ fixed}$$

and similarly, the conditional distribution on l is given by:

$$f(l|t) = \frac{f(l, t)}{\int f(l, t) \, \mathrm{d}l} \quad t \text{ fixed}$$

where the integrals are over the total range of the variables concerned. Bayes' theorem now gives:

$$f(l|t) \propto f(t|l) f(l).$$

As the total probability must be one, then

$$f(l|t) = \frac{f(t|l) f(l)}{\int f(t|l') f(l') \, \mathrm{d}l'}$$

as l is a continuous variable, and

$$f(x|y) = \frac{f(y|x) f(x)}{\sum_{x'} f(y|x') f(x')}$$

in the case where we are considering a discrete variable x, where the integral, or sum, is over all possible values of l' or x' respectively.

The example of the diagnostic system is a simple case where the variables could only take two values. The first is a yes/no for the presence of a fault, and the second yes/no for the diagnosis indicating a fault. Bayes' theorem can be demonstrated very simply using elementary algebra in this case, and this is left as an exercise for the reader.

In Bayesian statistics and data analysis, the two variables concerned are the data observed and the parameter(s) of interest. This means we have to consider a probability distribution of, for example, a failure rate, and before we can go into details, this must be dealt with, which is the subject of the next section.

6.3.4 Subjective probability

In earlier chapters on statistics, probability was introduced as a relative frequency. Common examples of probability are concerned with tossing coins and packs of cards etc., where an experiment can be repeated many times to test the notion that the probability of a given event has a particular value.

Probability is frequently used in other situations, that cannot be subjected to such a test. Consider the following questions, which could all be answered probabilistically.

1. Will it rain tomorrow?
2. Which political party will win the next general election?
3. Which is further north, Newcastle upon Tyne (UK) or Anchorage (USA)?
4. How old is the author's son?

Consider how the answers to these questions might be given in terms of probabilities.

1.   Data from past meteorological records can give an answer to this as a probability (in some parts of the world, weather forecast are given probabilistically), but if tomorrow happens to be important to you — your wedding day, for example — it is a one-off event, that will not be repeated, so the question then becomes 'will it rain on my wedding day?', and there is no way that you can carry out a trial, and have many wedding days!

2.   This question is one that is frequently asked, at the appropriate time! There is certainly only one 'next election', and it is not repeatable, although possibly events in the election could be identified that are repeatable, and the election broken down into small events that can be analysed using past data. This example is certainly bordering on the 'not testable' regime.

3.   This is certainly not testable! Either Newcastle is north of Anchorage or Anchorage is north of Newcastle, but, without reference to a map, which way would you be prepared to bet? What odds would you be prepared to accept or offer? Trying to answer these questions gives an idea of what your probability is. (And there are no right answers to these questions — only each individual's subjective idea, which depends on his or her experience.)

4.   This, like the previous question, is not testable. He is a fixed age. However, you might like to try putting a probability distribution on his age — which will be personal to you. Try thinking of a most likely age, and a minimum and maximum. Now suppose I told you he was a university student. How would that effect your distribution? He might

be one of these geniuses that went to university at a very young age, or he might have been a mature student, don't forget. Oh, and I should add, at the time of writing, he's a postgraduate. Does this influence your distribution?

These examples are designed to get across the idea of subjective probability and probability distributions, that reflect, not a long term relative frequency, but the uncertainty in the knowledge that we have about a system. The answer to the last two questions is a certainty—only it is unknown—and we can express that lack of knowledge using probabilities. In Bayesian analysis, experimenters and engineers put probability distributions on such things as failure rates and reliabilities, in order to express their uncertainty about their actual values. Classical statistics does not allow you to do this, as there is no way of collecting the data to verify the distributions given.

6.3.5 Bayesian analysis

Suppose we have some life data, $D, t_1, t_2, t_3, \ldots$ (where D is a set of times), and it is required to estimate the failure rate (which a classical statistician would consider constant). Then Bayes' theorem states:

$$P(\lambda|D) \propto P(D|\lambda)P(\lambda)$$

where:

$P(\lambda|D)$ is the probability density function on λ, the failure rate, given the data set above, called the posterior distribution.

$P(D|\lambda)$ is the probability of observing that particular data set, given λ. This is the likelihood function, defined in Section 5.2.2.

$P(\lambda)$ is the marginal distribution on λ, called the prior. It is often, although not always, an expression of the experimenter's lack of certainty about the actual value of λ. It is usually denoted $\Pi(\lambda)$.

This is just the application of Bayes' theorem to the two variables, the data and the parameter of interest, in this case λ, given that probability is interpreted to include the broader sense of an expression of lack of knowledge.

Suppose D comprises life data from an item which has a constant failure rate. Then the likelihood function is

$$\lambda^n e^{-\lambda T}$$

where n is the number of failures and T the total time on test, and the application of Bayes' theorem gives

$$P(\lambda|D) \propto \lambda^n e^{-\lambda T} \Pi(\lambda)$$

or

$$P(\lambda|D) = \frac{\lambda^n e^{-\lambda T}\Pi(\lambda)}{\int l^n e^{-lT}\Pi(l)\mathrm{d}l}$$

normalizing so that the area under the pdf is one. The integral is taken over all possible values of l, usually from 0 to ∞.

From the posterior it is possible to obtain information of a nature that one would obtain from a classical analysis. A point estimate of λ, for example, might be the mode of the posterior, the MLE, but it may be just as appropriate, depending on the circumstances, to use the expected value, or the median. The 95th percentile gives a value that has a 5 per cent probability of being exceeded, and corresponds to the 95 per cent upper confidence limit of classical statistics.

Bayesian intervals may be calculated, just as confidence intervals are given in classical statistics. The usual way is as follows. Suppose $P(\lambda)$ is a pdf for λ, a failure rate. Then the α confidence interval consists of those values of λ that lie between λ_0 and λ_1, such that

$$\int_{\lambda_0}^{\lambda_1} P(\lambda)\mathrm{d}\lambda = \alpha$$

and

$$P(\lambda_0) = P(\lambda_1).$$

This is illustrated in Fig. 6.25, where the area of the shaded portion under the curve is α.

6.3.6 Priors and conjugate priors

The prior, Π, is the experimenter's assessment of what value λ may be, expressed as a probability distribution. It may be subjective judgment alone, or it may be derived from previous data. The next section deals with a succession of trials, but if there are no design changes between trials, and nothing else to cause us to believe that the reliability changes from one set of trials to the next, then the posterior from one trial is the prior for the next, and this gives the same result as if the data had all been combined and the initial prior used.

Now let us look at the example of the previous section again. Suppose we had chosen a prior of the form

$$\Pi(\lambda) \propto \lambda^{\alpha-1}e^{-\beta\lambda}$$

i.e. λ is gamma distributed. Then applying Bayes' theorem gives

$$P(\lambda|D) \propto \lambda^{n+\alpha-1}e^{-(T+\beta)\lambda}$$

$$= \lambda^{\alpha'}e^{-\beta'\lambda}$$

$$= \lambda^{\alpha'-1}e^{-\beta'\lambda}$$

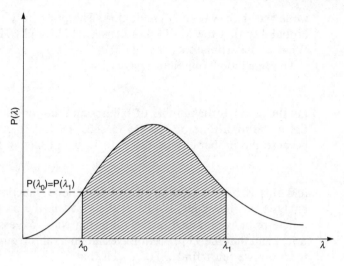

Fig. 6.25 Illustration of Bayesian interval.

where

$$\alpha' = \alpha + n$$

$$\beta' = \beta + T$$

and the posterior is of the same mathematical form as the prior. This has obvious advantages when performing calculations. Such a prior, chosen in such a way that the mathematical form remains the same, but with the values of the parameters of the pdf changed, is called a conjugate prior.

There is some debate in the Bayesian statistics community about the ethics of using conjugate priors, as they restrict the engineer in his choice of prior, but in practice, getting engineers to think along the right lines can be difficult, as they are not used to expressing their uncertainty as a distribution (expressing uncertainty as a probability is OK for a yes/no situation for most people—what is the probability that this system is of an acceptable reliability, for example, or what is the probability that a given design problem is solved). In practice the author has found that asking for the most likely value of the failure rate and the worst case failure rate, and using them as the mode and 95th percentile respectively gives acceptable results.

Examples 6.3 6.3.1 Six components were left running for 60 hr or until they had failed. At the end of this time, five had failed at times

$$15, \quad 23, \quad 27, \quad 31, \quad 45$$

while the sixth was still functioning. The prior information is that the probability that the MTTF lies between 15 hr and 60 hr is 95 per cent. What is the estimate of the MTTF?

The likelihood function is given by

$$\lambda^5 e^{-201\lambda}$$

i.e. the power is the number of failures and the constant in the exponential is the total time on test ($= 15 + 23 + 27 + 31 + 45 + 60$). This is because the probability of the sixth item not having failed by 60 hr is

$$e^{-60\lambda}$$

and this is its contribution to the likelihood function. Otherwise the analysis, to this point, is just as described in the previous example.

The prior is obtained in the following way. We assume a conjugate prior, and that the 95 per cent Bayesian interval is from 15 to 60 hr. That is to say we must find a $\Pi(\lambda)$ such that

$$\Pi(1/15) = \Pi(1/60)$$

and

$$\int_{1/15}^{1/60} \Pi(\lambda)\,d\lambda = 0.95.$$

As it is to be a conjugate prior, it can be shown that

$$\Pi(\lambda) \propto \lambda^{13} e^{-357\lambda}.$$

This gives the posterior

$$P(\lambda) \propto \lambda^{18} e^{-578\lambda}.$$

The graphs of the pdfs are shown in Figs 6.26 and 6.27. Also shown are the MLE, at $\lambda = 0.031$, the mean, at $\lambda = 0.033$, and the Bayesian 95 per cent interval, from 0.019 to 0.048. These correspond to an MLE of 32.3 hr, and a 95 per cent Bayesian interval from 20.8 hr to 52.6 hr.

6.3.2 Twenty components were tested to see if they functioned. One failed, and the other 19 functioned satisfactorily.

The likelihood function is given by

$$20R^{19}(1 - R)$$

and if we assume a prior of the form

$$\Pi(R) \propto R^{\alpha - 1}(1 - R)^{\beta - 1}$$

then this prior will be conjugate, as the posterior will be given by

$$P(R) \propto R^{\alpha + 19 - 1}(1 - R)^{\beta + 1 - 1} = R^{\alpha + 18}(1 - R)^{\beta}$$

i.e. R will be beta distributed. (See Section 2.5.6.) When estimating a reliability, the conjugate prior must be a beta distribution.

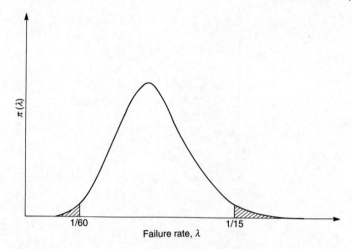

Fig. 6.26 Prior distribution of λ.

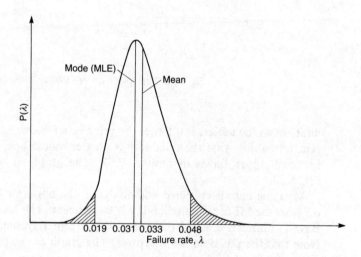

Fig. 6.27 Posterior distribution of λ.

For example, if

$$\alpha = 5.5$$

and

$$\beta = 1.5,$$

then π has a mode at 0.9 (i.e. the prior belief is that 90 per cent is the most likely value of the reliability), and the lower 95 per cent Bayesian

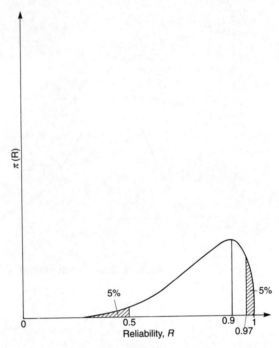

Fig. 6.28 First prior on reliability.

limit, from the tables, is 0.5 (i.e. the prior belief is that there is a 95 per cent probability that the reliability is greater than 50 per cent). The 95 per cent upper Bayesian limit is 0.97. The graph of π is shown in Fig. 6.28.

After the data is collected and analysed, the posterior has a mode at 0.94, so the MLE of the reliability is 94 per cent. The lower 95 per cent Bayesian limit is 0.8 and the upper 95 per cent Bayesian limit is 0.98. Note that the pdf becomes narrower. The graph of the posterior pdf is shown in Fig. 6.29.

6.3.7 Series of trials

If a series of trials is carried out, and there are no design changes between them, then using the posterior from one trial as the prior for the next gives the same result as would be obtained by amalgamating all the data and analysing it in one go using the first prior.

As an illustration, taking the second example above, suppose a further 30 components were examined, and two were found to be faulty. Then the likelihood function is

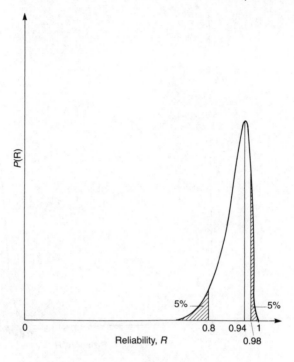

Fig. 6.29 First posterior and second prior on reliability.

$$C_2^{30} R^{28} (1 - R)^2$$

and using the posterior from the previous experiment as prior for the analysis in this case, gives the posterior

$$P(R) \propto R^{\alpha + 18 + 28}(1 - R)^{\beta + 2} = R^{\alpha + 46}(1 - R)^{\beta + 2}.$$

On the other hand, putting the results of the two trials together gives 50 trials, with three failures and 47 successes. This gives the likelihood function

$$C_3^{40} R^{47} (1 - R)^3$$

and using the original prior, gives the posterior

$$P(R) \propto R^{\alpha + 18 + 28}(1 - R)^{\beta + 2} = R^{\alpha + 46}(1 - R)^{\beta + 2}$$

which is what we obtained by analysing the two sets of data separately.

The posterior now has a mode at 0.934, a lower 95 per cent Bayesian limit at 0.84 and an upper 95 per cent Bayesian limit at 0.957. The pdf is shown in Fig. 6.30. Note that the pdf is becoming narrower as more data is collected (as would be expected).

These three sets of results can be summarised in Fig. 6.31, which shows the MLE and the three percentage points mentioned.

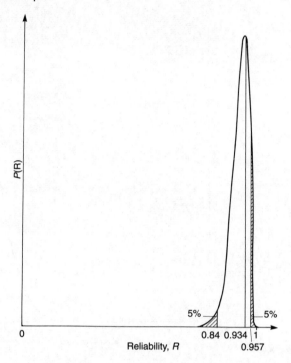

Fig. 6.30 Second posterior on reliability.

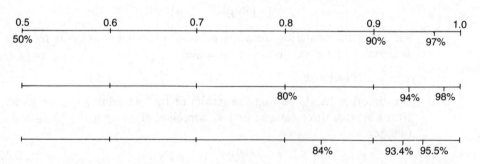

Fig. 6.31 Summary of Figs 6.28–6.30 showing percentage points and MLEs.

When analysing growth data, it is now possible to input the designer's judgement, and also deal with the problem of obtaining interval estimates, which is not possible with the FEF method. Suppose we were to examine the above data, but this time we shall assume that there was a design change between the two trials. To recap, the MLE of the reliability after the first trial was 94 per cent, and the lower 95 per cent Bayesian limit 80 per cent. A design change is incorporated that is supposed to improve the reliability. After the design change, the engineer believes

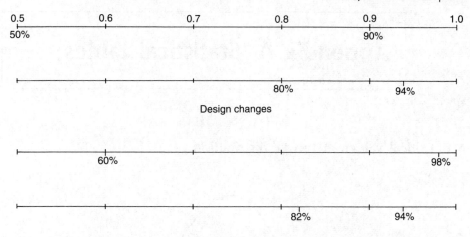

Fig. 6.32 Summary of reliability growth showing percentage points and MLEs.

the reliability is most likely to be 98 per cent, but he is less certain, and puts the lower 95 per cent limit at about 60 per cent. This gives values of α and β of 5.9 and 1.1 respectively.

After the data were collected and analysed, the new values of α and β were 33.2 and 3.1 respectively, which gives an MLE of 0.94 and a lower 95 per cent limit of 0.82.

These results are summarized in Fig. 6.32, which is similar to Fig. 6.31, in that it shows the way the MLE and lower limits change, but also shows the effects of the design change between the second and third trials.

Appendix A. Statistical tables

A.1 Critical values of the χ^2 distribution

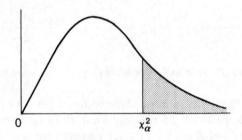

The table shows values of χ^2 for which

$$\int_{\chi^2}^{\infty} f(t)\, dt = \alpha$$

when $f(t)$ is the pdf of the χ^2 distribution.

ν	0.995	0.99	0.975	0.95	0.05	0.025	0.01	0.005
1	0.0^4393	0.0^3157	0.0^3982	0.0^2393	3.841	5.024	6.635	7.879
2	0.0100	0.0201	0.0506	0.103	5.991	7.378	9.210	10.597
3	0.0717	0.115	0.216	0.352	7.815	9.348	11.345	12.838
4	0.207	0.297	0.484	0.711	9.488	11.143	13.277	14.860
5	0.412	0.554	0.831	1.145	11.070	12.832	15.086	16.750
6	0.676	0.872	1.237	1.635	12.592	14.449	16.812	18.548
7	0.989	1.239	1.690	2.167	14.067	16.013	18.475	20.278
8	1.344	1.646	2.180	2.733	15.507	17.535	20.090	21.955
9	1.735	2.088	2.700	3.325	16.919	19.023	21.666	23.589
10	2.156	2.558	3.247	3.940	18.307	20.483	23.209	25.188
11	2.603	3.053	3.816	4.575	19.675	21.920	24.725	26.757
12	3.074	3.571	4.404	5.226	21.026	23.337	26.217	28.300
13	3.565	4.107	5.009	5.892	22.362	24.736	27.688	29.819
14	4.075	4.660	5.629	6.571	23.685	26.119	29.141	31.319
15	4.601	5.229	6.262	7.261	24.996	27.488	30.578	32.801

				α				
ν	0.995	0.99	0.975	0.95	0.05	0.025	0.01	0.005
16	5.142	5.812	6.908	7.962	26.296	28.845	32.000	34.267
17	5.697	6.408	7.564	8.672	27.587	30.191	33.409	35.718
18	6.265	7.015	8.231	9.390	28.869	31.526	34.805	37.156
19	6.844	7.633	8.907	10.117	30.144	32.852	36.191	38.582
20	7.434	8.260	9.591	10.851	31.410	34.170	37.566	39.997
21	8.034	8.897	10.283	11.591	32.671	35.479	38.932	41.401
22	8.643	9.542	10.982	12.338	33.924	36.781	40.289	42.796
23	9.260	10.196	11.689	13.091	35.172	38.076	41.638	44.181
24	9.886	10.856	12.401	13.848	36.415	39.364	42.980	45.558
25	10.520	11.524	13.120	14.611	37.652	40.646	44.314	46.928
26	11.160	12.198	13.844	15.379	38.885	41.923	45.642	48.290
27	11.808	12.879	14.573	16.151	40.113	43.194	46.963	49.645
28	12.461	13.565	15.308	16.928	41.337	44.461	48.278	50.993
29	13.121	14.256	16.047	17.708	42.557	45.722	49.588	52.336
30	13.787	14.953	16.791	18.493	43.773	46.979	50.892	53.672

* Abridged from Table 8 of *Biometrika Tables for Statisticians*, Vol. I, by permission of E. S. Pearson and the Biometrika Trustees.

A.2 Kolmogorov–Smirnov tables

Critical values of the maximum absolute difference between sample $F_n(x)$ and population $F(x)$ cumulative distribution.

Number of trials, n	Level of significance, α			
	0.10	0.05	0.02	0.01
1	0.95000	0.97500	0.99000	0.99500
2	0.77639	0.84189	0.90000	0.92929
3	0.63604	0.70760	0.78456	0.82900
4	0.56522	0.62394	0.68887	0.73424
5	0.50945	0.56328	0.62718	0.66853
6	0.46799	0.51926	0.57741	0.61661
7	0.43607	0.48342	0.53844	0.57581
8	0.40962	0.45427	0.50654	0.54179
9	0.38746	0.43001	0.47960	0.51332
10	0.36866	0.40925	0.45662	0.48893
11	0.35242	0.39122	0.43670	0.46770
12	0.33815	0.37543	0.41918	0.44905
13	0.32549	0.36143	0.40362	0.43247
14	0.31417	0.34890	0.38970	0.41762
15	0.30397	0.33760	0.37713	0.40420
16	0.29472	0.32733	0.36571	0,39201
17	0.28627	0.31796	0.35528	0.38086
18	0.27851	0.30936	0.34569	0.37062
19	0.27136	0.30143	0.33685	0.36117
20	0.26473	0.29408	0.32866	0.35241
21	0.25858	0.28724	0.32104	0.34427
22	0.25283	0.28087	0.31394	0.33666
23	0.24746	0.27490	0.30728	0.32954
24	0.24242	0.26931	0.30104	0.32286
25	0.23768	0.26404	0.29516	0.31657
26	0.23320	0.25907	0.28962	0.31064
27	0.22898	0.25438	0.28438	0.30502
28	0.22497	0.24993	0.27942	0.29971
29	0.22117	0.24571	0.27471	0.29466
30	0.21756	0.24170	0.27023	0.28987
31	0.21412	0.23788	0.26596	0.28530
32	0.21085	0.23424	0.26189	0.28094
33	0.20771	0.23076	0.25801	0.27677
34	0.20472	0.22743	0.25429	0.27279
35	0.20185	0.22425	0.26073	0.26897
36	0.19910	0.22119	0.24732	0.26532
37	0.19646	0.21826	0.24404	0.26180
38	0.19392	0.21544	0.24089	0.25843

Number of trials, n	Level of significance, α			
	0.10	0.05	0.02	0.01
39	0.19148	0.21273	0.23786	0.25518
40[b]	0.18913	0.21012	0.23494	0.25205

$N > 40 \approx \frac{1.22}{N^{1/2}}, \frac{1.36}{N^{1/2}}, \frac{1.51}{N^{1/2}}$ and $\frac{1.63}{N^{1/2}}$ for the four levels of significance.

A.3 Critical values of the *t* distribution

The table shows values of t for which

$$\int_t^\infty f(x)\,\mathrm{d}x = \alpha$$

when $f(x)$ is the pdf of the t distribution.

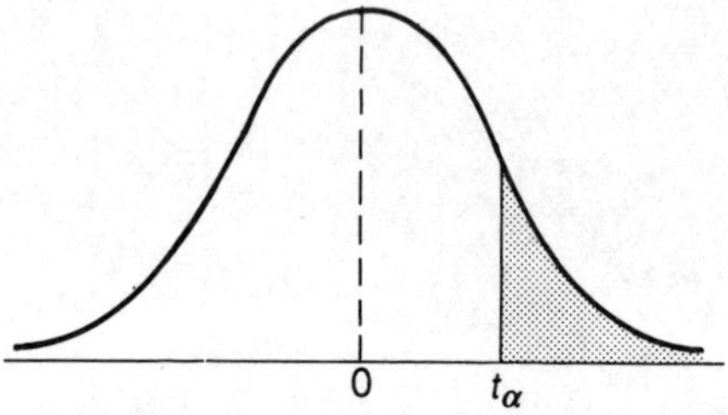

	α				
ν	0.10	0.05	0.025	0.10	0.005
1	3.078	6.314	12.706	31.821	63.657
2	1.886	2.920	4.303	6.965	9.925
3	1.638	2.353	3.182	4.541	5.841
4	1.533	2.132	2.776	3.747	4.604
5	1.476	2.015	2.571	3.365	4.032
6	1.440	1.943	2.447	3.143	3.707
7	1.415	1.895	2.365	2.998	3.499
8	1.397	1.860	2.306	2.986	3.355
9	1.383	1.833	2.262	2.821	3.250
10	1.372	1.812	2.228	2.764	3.169
11	1.363	1.796	2.201	2.718	3.106
12	1.356	1.782	2.179	2.681	3.055
13	1.350	1.771	2.160	2.650	3.012
14	1.345	1.761	2.145	2.624	2.977
15	1.341	1.753	2.131	2.602	2.947
16	1.337	1.746	2.120	2.583	2.921
17	1.333	1.740	2.110	2.567	2.898
18	1.330	1.734	2.101	2.552	2.878

ν	α				
	0.10	0.05	0.025	0.10	0.005
19	1.328	1.729	2.093	2.539	2.861
20	1.325	1.725	2.086	2.528	2.845
21	1.323	1.721	2.080	2.518	2.831
22	1.321	1.717	2.074	2.508	2.819
23	1.319	1.714	2.069	2.500	2.807
24	1.318	1.711	2.064	2.492	2.797
25	1.316	1.708	2.060	2.485	2.787
26	1.315	1.706	2.056	2.479	2.779
27	1.314	1.703	2.052	2.473	2.771
28	1.313	1.701	2.048	2.467	2.763
29	1.311	1.699	2.045	2.462	2.756
inf.	1.282	1.645	1.960	2.326	2.576

*Table A.5 is taken from Table IV of R. A. Fisher: *Statistical Methods for Research Workers*, published by Oliver & Boyd Ltd., Edinburgh, by permission of the author and publishers.

A.4 Critical values of the β distribution

The table shows values of R for which

$$\int_R^1 Kr^{\alpha-1}(1-r)^{\beta-1}\,dr = 0.05$$

where K is the normalizing constant

α \ β	0.5	1.0	1.5	2.0	2.5	3.0	3.5	4.0	4.5
0.5	.0061558	.0025000	.0015429	.0011119	.0^{3}008620	.0^{3}71179	.0^{3}60300	.0^{3}52300	.0^{3}46170
1.0	.097500	.050000	.033617	.025321	.020308	.016952	.014548	.012741	.011334
1.5	.22852	.13572	.097308	.076010	.062413	.052962	.046007	.040671	.036447
2.0	.34163	.22361	.16825	.13535	.11338	.097611	.085727	.076440	.068979
2.5	.43074	.30171	.23553	.19403	.16528	.14409	.12778	.11482	.10427
3.0	.50053	.36840	.29599	.24860	.21477	.18926	.16927	.15316	.13989
3.5	.55593	.42489	.34929	.29811	.26063	.23182	.20890	.19019	.17461
4.0	.60071	.47287	.39607	.34259	.30260	.27134	.24613	.22532	.20783
4.5	.63751	.51390	.43716	.38245	.34080	.30777	.28082	.25835	.23930
5.0	.66824	.54928	.47338	.41820	.37553	.34126	.31301	.28924	.26894
5.5	.69425	.58003	.50546	.45033	.40712	.37203	.34283	.31807	.29677
6.0	.71654	.60696	.53402	.47930	.43590	.40031	.37044	.34494	.32286
6.5	.73583	.63073	.55958	.50551	.46219	.42635	.39604	.37000	.34732
7.0	.75268	.65184	.58256	.52932	.48626	.45036	.41980	.39338	.37025
7.5	.76754	.67070	.60333	.55102	.50836	.47255	.44187	.41521	.39176
8.0	.78072	.68766	.62217	.57086	.52872	.49310	.46242	.43563	.41196
8.5	.79249	.70297	.63933	.58907	.54750	.51217	.48159	.45474	.43094
9.0	.80307	.71687	.65503	.60584	.56490	.52991	.49949	.47267	.44880
9.5	.81263	.72954	.66944	.62131	.58103	.54645	.51624	.48951	.46564
10.0	.82131	.74113	.68271	.63564	.59605	.56189	.53194	.50535	.48152
10.5	.82923	.75178	.69496	.64894	.61004	.57635	.54669	.52027	.49652
11.0	.83647	.76160	.70632	.66132	.62312	.58990	.56056	.53434	.51071
11.5	.84313	.77067	.71687	.67287	.63536	.60263	.57363	.54764	.52415
12.0	.84927	.77908	.72669	.68366	.64684	.61461	.58596	.56022	.53689
12.5	.85494	.78690	.73586	.69377	.65764	.62590	.59761	.57213	.54898
13.0	.86021	.79418	.74444	.70327	.66780	.63656	.60864	.58343	.56048
13.5	.86511	.80099	.75249	.71219	.67738	.64663	.61909	.59416	.57141
14.0	.86967	.80736	.76004	.72060	.68643	.65617	.62900	.60436	.58183
14.5	.87394	.81334	.76715	.72854	.69499	.66522	.63842	.61407	.59177
15.0	.87794	.81896	.77386	.73604	.70311	.67381	.64738	.62332	.60125
20.0	.90734	.86089	.82447	.79327	.76559	.74053	.71758	.69636	.67663
30.0	.93748	.90497	.87881	.85591	.83517	.81606	.79824	.78150	.76569
60.0	.96837	.95130	.93720	.92458	.91290	.90192	.89148	.88150	.87191
∞	1.00000	1.00000	1.00000	1.00000	1.00000	1.00000	1.00000	1.00000	1.00000

α \ β	5.0	6.0	7.5	10.0	12.0	15.0	20.0	30.0	60.0
0.5	$.0^3 41325$	$.0^3 34154$	$.0^3 27098$	.20156	$.0^3 16727$	$.0^3 13326$	$.0^4 99535$	$.0^4 66082$	$.0^4 32904$
1.0	.010206	.0085124	.0068158	.0051162	.0042653	.0034137	.0025614	.0017083	$.0^2 85452$
1.5	.033020	.027794	.022465	.017026	.014264	.011472	.0086511	.0057991	.0029157
2.0	.062850	.053376	.043541	.033319	.028053	.022679	.017191	.011585	.0058568
2.5	.095510	.081790	.067312	.051995	.043994	.035747	.027240	.018458	.0093841
3.0	.12876	.11111	.092207	.071870	.061103	.049898	.038224	.026043	.013317
3.5	.16142	.14029	.11733	.092238	.078783	.064651	.049781	.034103	.017540
4.0	.19290	.16875	.14216	.11267	.096658	.079695	.061676	.042481	.021976
4.5	.22292	.19618	.16638	.13288	.11449	.094827	.073748	.051068	.026572
5.0	.25137	.22244	.18984	.15272	.13211	.10991	.085885	.059786	.031288
5.5	.27823	.24746	.21244	.17207	.14943	.12484	.098008	.068575	.036094
6.0	.30354	.27125	.23413	.19086	.16636	.13955	.11006	.077394	.040967
6.5	.32737	.29383	.25492	.20908	.18288	.15401	.12199	.086209	.045889
7.0	.34981	.31524	.27481	.22669	.19895	.16818	.13377	.094994	.050847
7.5	.37095	.33554	.29382	.24370	.21457	.18203	.14539	.10373	.055827
8.0	.39086	.35480	.31199	.26011	.22972	.19556	.15682	.11240	.060821
8.5	.40965	.37307	.32936	.27594	.24441	.20877	.16805	.12099	.065820
9.0	.42738	.39041	.34596	.29120	.25865	.22164	.17908	.12950	.070818
9.5	.44414	.40689	.36183	.30591	.27244	.23418	.18989	.13791	.075809
10.0	.45999	.42256	.37701	.32009	.28580	.24639	.20050	.14622	.080789
10.5	.47501	.43746	.39154	.33375	.29874	.25828	.21088	.15442	.085753
11.0	.48925	.45165	.40544	.34693	.31126	.26985	.22106	.16252	.090698
11.5	.50276	.46518	.41877	.35964	.32340	.28112	.23102	.17051	.095622
12.0	.51560	.47808	.43154	.37190	.33515	.29208	.24077	.17838	.10052
12.5	.52782	.49040	.44379	.38373	.34653	.30275	.25032	.18615	.10539
13.0	.53945	.50217	.45554	.39516	.35756	.31314	.25966	.19379	.11024
13.5	.55054	.51343	.46683	.40619	.36826	.32325	.26880	.20133	.11505
14.0	.56112	.52420	.47768	.41685	.37862	.33309	.27775	.20875	.11983
14.5	.57122	.53452	.48812	.42715	.38867	.34267	.28650	.21606	.12458
15.0	.58088	.54442	.49816	.43711	.39842	.35200	.29507	.22326	.12930
20.0	.65819	.62459	.58083	.52099	.48175	.43321	.37136	.28936	.17453
30.0	.75070	.72282	.68535	.63185	.59522	.54807	.48477	.39458	.25416
60.0	.86266	.84504	.82047	.78342	.75661	.72016	.66738	.58326	.42519
∞	1.00000	1.00000	1.00000	1.00000	1.00000	1.00000	1.00000	1.00000	1.00000

Appendix B. Solutions to selected exercises

Exercises 1.1

1.1.1. This exercise is very open ended, and the answer will depend on the students' own experience, and to a certain extent to where they live. So the following are for guidance only, but should all be considered.

(a) Freezer

(i) Environment
Temperature and humidity. These can be extreme, as freezers are kept in all sorts of strange places, as well as the kitchen (garage, shed, etc.) and so temperature could range from $-10°C$ to $80°C$ or more extreme. Similarly the humidity could be from very dry to 100%.

The user is untrained, and there is little or no maintenance, except for being cleaned annually.

Other aspects include physical knocks from children etc., a certain amount of pollution (exhaust fumes for those that are kept in a garage, for example), including grease (in the kitchen).

(ii) Mission
Primarily, to keep food at a low temperature, within certain limits, (both quantified) continuously.

A secondary functioning is that it should be easy to use, i.e. it should have a light, and the strength and size of the user must be considered.

(b) Fire extinguisher

(i) Environment
Temperature and humidity. These can be extreme in a warehouse, as there may be no heating or air conditioning, but not as bad as outside. They should be quantified. The temperature should there be a fire can be extremely hot (specify!).

The user receives minimum training, and there is little or no maintenance, except for an annual inspection.

Other aspects include pollutants, in the form of dust, fumes, and liquids (what might be found in a warehouse?, but it could include car exhaust from materials handling vehicles), and physical knocks.

(ii) The primary mission is to hold specified chemicals, inertly, until it is required to deliver a specified substance, at a specified rate and pressure, over a specified range, for a specified time.

Further consideration must be given to ease of use and weight etc. In particular it must be very reliable, in both operating modes (inert and use).

(c)   Floodlight

(i)   Environment
All weather, which means specifying the temperature extremes, humidity, rain, snow and ice (which adds to the loads on the structure), and wind (which also adds to the loads). These should be quantified, and note may be taken of the inability to state the maximum values for some of these, hence introducing an element of risk to the decisions involved in the design.

The user may or may not be trained. The way a floodlight is used allows time for maintenance, which may be by the user, or may be by the designer (depends on the contract).

Other aspects include pollutants in the atmosphere, and people, possibly attempting to climb the structure.

(ii)   The mission is to illuminate a sports ground for a specified length of time, possibly every evening.

1.1.2.   There should be at least four categories, namely:

A.   Safety related failures, i.e. those failures that could cause loss of life or serious injury.

B.   Total failure, i.e. those failures that prevent continuation of the mission.

C.   Those failures that cause impairment of the mission, or lead to a degraded functioning.

D.   Other failures.

There may be cause for more categories; for example, C could be replaced by the two categories:

C1.   Those failures that lead to degraded functioning, with a high risk of total failure.

C2.   Those failures that lead to degraded functioning, but the probability of total failure is low.

In any case there should not be too many categories, otherwise the system breaks down under its own weight. Four is a minimum, about six is probably ideal, and ten is a maximum (and systems of ten categories do exist in practice).

The failures in the exercise fall into the following categories:

brake failure A

courtesy light D

silencer C

pump B

puncture D (although this is more serious than the courtesy light. Maybe there is room for D to be subdivided.)

cannot change wheel B

1.1.3.   (a) Years. The mission time is its life, which may be several years (the author had a freezer which lasted 18 years).

(b) In the dormant state, months or years. In this case the mission is its life. In the active state, a few minutes (specified). A fire extinguisher lasts years.

(c) In the active state, hours. The mission could be two or three hours in UK. Its life is several years.

1.1.4. There are two functions, to give a signal in the event of a fire, and to remain silent otherwise (no false alarms). Detection reliability must be very high, depending on the consequences of the fire, the probability of failure must be the order of 10^{-6}, but the probability of a false alarm must be an order of magnitude lower than the probability of a fire, otherwise when there is a real fire, the alarm will not have the desired effect.

Exercises 1.2

1.2.1. The completed table is (solution in bold):

system	1	2	3	4	5	6
MTBF	**100**	150	**487**	**995**	67	200
λ	0.01	6.7×10^{-3}	2.1×10^{-3}	1×10^{-2}	1.5×10^{-2}	**0.005**
mission	10	20	25	10	15	**21.1**
R	**0.90**	**0.88**	0.95	0.99	0.8	0.9

1.2.2. MTTF = 41.8 hr, $R(20) = 62\%$

1.2.3. MTTF = 150 hr, $R(30) = 82\%$

1.2.4. R = proportion of items still functioning,

$= 0.8$ for Question 2

$= 0.9$ for Question 3.

Exercises 1.3

1.3.1. (a) Spare wheel, jack, spanner of the correct size.
Remove wheel, jack and spanner from car, check spare is functional, loosen nuts on wheel, jack up car, remove nuts on wheel, remove wheel, replace with spare, hand tighten nuts, lower jack, tighten nuts, replace wheel, jack and spanner in car.

(b) Spare fuze, screwdrivers of correct size.
Remove plug from wall, unscrew cover, remove old fuze and replace, replace cover, tighten screw, test by plugging in and turning on appliance.

(c) New washer, spanners.
Turn water off at main, drain appropriate part of system, disassemble tap, remove old washer, replace with new, reassemble tap, close tap, turn water on again, bleed air from system if necessary, test tap.

1.3.2. (a) Outdoors, all weathers. Individual may be untrained and not necessarily strong. Resources supplied by car manufacturer.

(b) Indoors, individual untrained, resources from amateur toolbox.

(c) As (b).

Exercises 1.4

1.4.1.   In a car ignition system, the components can and do degrade.
The functioning of the system is a function of the total degradation of the system,
and is very difficult to model, particularly as the weather may effect the system.
(Try starting a badly maintained car on a cold, damp March morning, for example.)

1.4.2. $R = 0.96(1135)$

1.4.3.   For system level redundancy, $R = 0.9985$

For component level redundancy, $R = 0.9994$

1.4.4.   For the ith component,

$$R_i = \exp(-\lambda t_i)$$

which gives the following:

i	1	2	3	4
λ_i	0.0051	0.016	0.011	0.002

and

$$R_s = \exp(-0.126t) + \exp(-0.234t) - \exp(-0.339t)$$

Exercises 2.1

2.1.1.   0.84

2.1.2.   (a) 0.726 75
   (b) 0.247 25
   (c) 0.025 23
   (d) 0.999 25
   (e) 0.000 75

2.1.3.   0.982 08

Exercises 2.2

2.2.1.

n	0	1	2	3
P_n	0.000 75	0.025 23	0.247 25	0.726 75

2.2.2.

n	0	1	2	3	total
P_n	0.000 125	0.007 125	0.135 375	0.857 357	1.0

2.2.3.

n	0	1	2	3	4	total
P_n	0.0001	0.0036	0.486	0.2916	0.6561	1.0

$$f(t) = 3t^2 e^{-t^2}, \quad \lambda(t) = 3t^2$$

Exercises 2.3

2.3.1. Mean = 253.6, median = 147, mean deviation from the mean = 43, from the median = 43, variance = 2490, sd = 48.9.

2.3.2. Mean = 5.14, median = 55.3, mean deviation from the mean = 14.9, from the median = 14.8, variance = 370, sd = 19.3, mode = 55.5.

2.3.4. MTTF = $2/\lambda$

$$R(t) = \lambda t e^{-\lambda t} + e^{-\lambda t}$$

Exercises 2.4

2.4.1. 0.7%

2.4.2. 98.5%

2.4.3. 3.6%

2.4.4. 2.5, 11%

2.4.5. 10%

2.4.6. 8.6%

2.4.7. 0.007

2.4.8. 39%

Exercises 4.1

4.1.1. $\alpha = 0.41$. 140 hr. 200 hr. 4670 hr. (by eye)

4.1.2. Something funny happens round about 1000 hr. This is borne out by the $n - T$ plot. Replotting the data after 1000 hr, after subtracting 1000 hr from each of the last 7 data, and 6 from n, the plot looks much better. $\alpha = 0.49$, $M_i = 236$ hr.

Exercises 4.2

4.2.1. 0.998, 0.977

4.2.2. UQL, LQL on m are 1.37, 7.75, corresponding to an MTTF of 730 hr, 129 hr.

4.2.3. Growth and other development data; good quality control procedures; an in service demo and/or growth warranty, though if the equipment is safety related, and only to be used in an emergency, the latter is not applicable.

Exercises 4.3

4.3.1. $\hat{\beta} = 1.25,\ \hat{\eta} = 780\ hr.$

4.3.2. $\hat{t}_0 = 202\ \text{hr},\ \hat{\beta} = 1.2,\ \hat{\eta} = 92\ \text{hr}.$

4.3.3. $\hat{\beta} = 2.0,\ \hat{\eta} = 85.$

Exercises 5.1

5.1.1. P(3 or more) = 1.4%. Significant at the 5% level, not at 1%. The null hypothesis that $R = 99\%$ is doubtful.

5.1.2. P(4 or more breakdowns) = 18%. Not at all significant, very little evidence to disbelieve the hypothesis.

5.1.3. χ^2 observed = 8.33 (= 2 × 1000/240) χ^2 5% = 6.57 < χ^2 observed, so not significant at the 5% level. Very little evidence to disbelieve the hypothesis.

Exercises 5.2

5.2.1. (96.8%, 99.4%)

5.2.2. (115 hr, 619 hr)

5.2.3. 675 hr.

Exercises 5.3

5.3.1. χ^2 observed = 1.13 < χ^2, $v = 2$, = 5.99. Not at all significant.

5.3.2. χ^2 observed = 0.9 < χ^2, $v = 1$, = 3.84, not significant.

5.3.3. From Weibull analysis, $\hat{\beta} = 2.1, \hat{\eta} = 94$. Group the data into intervals of length 40, $\chi^2 = 0.33 < \chi^2$, $v = 2$, = 3.84. There is very little evidence to suggest the data is not from a Weibull distribution.

5.3.4. Observed maximum difference = 0.1180. From the table, the value at the 10% level of significance is 0.468. The result is not at all significant.

5.3.5. From a Weibull plot, $\hat{\beta} = 0.8, \hat{\eta} = 98$. Observed maximum difference = 0.14, tabulated value is 0.468. The result is not significant.

Exercises 5.4

5.4.1. χ^2 observed = 0.32 < χ^2, $v = 1$, = 3.84. The result is not significant.

5.4.2. χ^2 observed $= 2.06 < \chi^2$, $v = 3$, $= 7.81$. The result is not significant.

5.4.3. χ^2 observed $= \chi^2$, $v = 1$, $= 3.84$. The result is not significant.

Exercise 5.5

If x is power output and y is lifetime, then

$$y = 106.2 - 0.078x$$

95% confidence intervals are

$$106.2 \pm 4.2 = (102.0, 110.4)$$
$$0.078 \pm 0.021 = (0.057, 0.099)$$
$$r = -0.95.$$

Exercises 6.1

6.1.2. Other possible policies are, 2 repair teams that can each deal with either sub-system, the system is only repaired when it fails totally, in which case it is repaired to the fully functioning state. In the latter case all the arrows representing repair (there are three of them) go direct to the state representing everything functions.

6.1.5. The differences are due to the different repair policies. The RBD approach assumes the failures and repairs are independent, which is not true unless there are sufficient repair teams and the system does not stop at system failure.

Bibliography

Books and journal papers

Barlow, R. E. and Proschan, F. (1981). Statistical theory of reliability and life testing. To Begin With, Silver Spring, MD, USA.

Berger, J. O. (1985). Statistical decision theory and Bayesian analysis, (2nd edn) Springer Verlag, New York.

Carter, A. D. S. (1986). Mechanical reliability, (2nd edn) Macmillan, London.

Cox, D. R. (1967). Renewal theory. Chapman and Hall, London.

Crowder, M. J., Kimber, A. C., Smith, R. L., and Sweeting, T. J. (1991). Statistical analysis of reliability data. Chapman and Hall, London.

DeGroot, M. H. (1986). Probability and statistics, (2nd ed.) Addison–Wesley.

Duane, J. T. (1965). Learning curve approach to reliability monitoring. IEEE transactions on aerospace, vol. 2, p. 563.

Grosh, D. L. (1989). A primer of reliability theory. Wiley, New York.

Kalplfleisch, J. and Prentice, R. L. (1980). The statistical analysis of failure time data. Wiley, New York.

Leitch, R. D. (1990). A statistical model of rough loading, Proceedings of the 7th international conference on reliability and maintainability, Brest, France.

Mann, N. R., Schafer, R. E., and Singpurwalla, N. D. (1974). Methods for statistical analysis of reliability and life data. Wiley, New York.

O'Connor, P. D. T. (1991). Practical reliability engineering, (3rd edn). Wiley, Chichester.

Ott, L. (1988). An introduction to statistical methods and data analysis, (3rd edn). PWS-Kent, Boston.

Pagés, A. and Gondran, M. (1986). System reliability, evaluation and prediction in engineering, (trans. E. Griffin). North Oxford Academic Publishers Ltd, London.

Petroski, H. (1985). To engineer is human. Macmillan, London.

Pile, S. (1979). The book of heroic failures. Futura Publications, London.

Sander, P. and Badoux, R. (Editors), (1991). Bayesian methods in reliability. Kluwer, Netherlands.

Villemeur, A. (1991). Reliability, availability, maintainability and safety assessment, vol. 1, (trans. A. Cartier and M-C. Lartisien). Wiley, Chichester.

White, J. S. (1967). The moments of log-Weibull order statistics. Report issued by General Motors Corporation Research Laboratories.

Standards

BS 5760: Reliability of systems, equipments and components, British Standards Institution, London.

BS 6001: Sampling procedures for inspection by attributes. British Standards Institution, London.

UK Defence Standard 00–40: The management of reliability and maintainability. HMSO.

UK Defence Standard 00–41/Issue 3: Reliability and maintainability MOD guide to practices and procedures. HMSO.

US MIL-STD 105D: Sampling procedures and tables for inspection by attributes. Available from the National Technical Information Service, Springfield, Virginia.

US MIL-HDBK 217: Reliability prediction for electronic systems. Available from the National Technical Information Service, Springfield, Virginia.

US MIL-STD 781: Reliability testing for equipment development, qualification and production. Available from the National Technical Information Service, Springfield, Virginia.

US MIL-STD 1629: Failure mode and effects analysis. Available from the National Technical Information Service, Springfield, Virginia.

Index